STUDENT'S STUDY MANUAL
to accompany *CAROLA, HARLEY, AND NOBACK*
Human Anatomy

STUDENT'S STUDY MANUAL

to accompany *CAROLA, HARLEY, AND NOBACK*

Human Anatomy

JOHN P. HARLEY

Eastern Kentucky University

McGraw-Hill, Inc.

New York   St. Louis   San Francisco   Auckland   Bogotá   Caracas
Lisbon   London   Madrid   Mexico   Milan   Montreal   New Delhi
Paris   San Juan   Singapore   Sydney   Tokyo   Toronto

STUDENT'S STUDY MANUAL
to accompany Carola, Harley, and Noback
Human Anatomy

Copyright © 1992 by McGraw-Hill, Inc. All rights reserved.
Printed in the United States of America.
Except as permitted under the United States Copyright Act of 1976,
no part of this publication may be reproduced or distributed
in any form or by any means, or stored in a data base or retrieval system,
without the prior written permission of the publisher.

1 2 3 4 5 6 7 8 9 0 MAL MAL 9 0 9 8 7 6 5 4 3 2 1

ISBN 0-07-010547-2

This book was set in Times Roman and Helvetica typefaces by Strong House, Inc.
The editor was Kathi M. Prancan;
the art director was Gayle Jaeger;
the production supervisor was Janelle S. Travers.
Malloy Lithographing, Inc., was printer and binder.

Photo credits: Joel Gordon, page 22;
Custom Medical Stock Photo, pages 20, 21, 23, 24, 25, 26, 27, and 28.

Contents

Preface

This ***Student's Study Manual*** accompanies *Human Anatomy* by Robert Carola, John P. Harley, and Charles R. Noback. It is designed to help the student master the many areas of human anatomy. Throughout the manual there is extensive cross-referencing by page number, making it easy to refer to the textbook's discussion of specific topics. Anatomical language relevant to the student's experience is used to make him/her comfortable with the study of the human body.

The chapters of the study manual correspond to those of the text. Each chapter of the manual contains several parts:
1 Prefixes and Suffixes
2 Developing Your Outline
3 Major Terms
4 Boxes
5 Labeling Activity
6 Post-Test

The **Prefixes and Suffixes** section contains morphemes (any of the minimal grammatical units of the language of anatomy that cannot be divided into smaller parts). Students will find it easier to master the vocabulary of anatomy if they progressively learn these word parts.

The **Developing Your Outline** section includes a variety of study activities that identify and review the principal ideas of each chapter. This section follows an outline of the chapter, using the same major headings and subheadings as the text. Answering the questions while reading the chapter can be a very effective learning tool. A completed Developing Your Outline section acts as a chapter summary and will be useful for review. The thought-provoking exercises in this section are designed to increase the student's depth of understanding of the textbook material and, overall, of the human body.

Major Terms lists all of the important terms from the text. The student can complete the list by defining or describing the term in the space beside it. The completed lists can be used as a comprehensive overview of terminology, as a guide to examinations, to gain an appreciation for the flow of text material as well as the volume of material that is new within each chapter. These terms will be used repeatedly in the textbook and in the different sections of this manual. In many cases, entries in one chapter are repeated in subsequent chapters since they are integral components of the study of human anatomy.

The **Boxes** found in most chapters describe selected careers in the allied health professions that the student may find interesting. The careers relate to material discussed in the chapter.

The **Labeling Activity** section presents unlabeled illustrations from the textbook and gives the student an opportunity to label diagrams based on the terminology in the chapter.

The **Post-Test** contains questions for each chapter that reinforce the learning objectives, text material, and certain operational skills. An answer key to the post-test is provided at the end of the manual to give immediate feedback about comprehension. The answers should be checked only after all questions have been completed for the chapter. It will then be immediately apparent if there are any areas that require further study or review.

This *Student's Study Manual* is a self-help tool. It can assist in learning human anatomy and evaluating the student's knowledge of this science only to the extent that it is used correctly and regularly.

Finally, please let us know about any inaccuracies or suggestions for improvement in this manual. Contact the Biology Editor at McGraw-Hill Publishing Company.

John P. Harley, Ph.D.

To the Student

One of the most important factors contributing to a student's success in college, and in a course in anatomy, is the development of a good study technique. This study manual has been written and designed to help you develop that technique. To ensure that you will obtain the most from this textbook and your course, we would like to offer some suggestions to help you study and learn as effectively as possible.

TIME MANAGEMENT AND THE STUDY ENVIRONMENT

Anatomy courses cover much detailed information over the semester. Keep this in mind when you are planning your study time for this course. Managing your time well is one factor in your success.

A proper study environment will help you concentrate properly. Try to find a quiet place with a desk and good lighting. If possible, always study in the same place, and use it only for studying. This approach will help condition you to be mentally alert and ready to study.

MAKING THE MOST OF LECTURES AND LABORATORIES

One important way to keep up with an anatomy course is to attend all lectures and laboratory sessions. To get the most from your classes, read the relevant textbook material beforehand, and be prepared to take an active part. Ask questions when something isn't clear—the chances are good that other students don't understand it either. Once you start asking questions, other students will join in.

During a lecture take notes clearly so that you can understand them later. Try to capture the main ideas, concepts, and important facts and definitions, especially if the lecturer extends the discussion of a specific topic beyond what is in in the textbook. As soon as possible after a lecture, review your notes to be sure that they are complete and that you understand them. Refer to your textbook if you are uncertain about something in your notes, and emphasize important points in your textbook by underlining or highlighting.

GETTING THE MOST FROM YOUR TEXTBOOK

Your textbook can be an important learning tool. There are many ways to use your textbook, and you may have already developed a technique in other courses. Or, you may want to try the following technique, called SQ4R (survey, question, read, revise, record, and review):*

1. Survey. Briefly scan the chapter to familiarize yourself with its general content. Read the title, introduction, summary, and main headings. Make a list of the major points you think the chapter will cover. Pay close attention to a chapter outline and chapter objectives if the text has these features.

* L. L. Thistlethwaite and N. K. Snouffer, *College Reading Power,* 3d ed. Dubuque, IA: Kendall/ Hunt, 1976.

2. Question. As you read each main heading or subheading, compose a general question or two that you believe the section will answer. This preview question should help you focus your reading of the particular section. Ask yourself more questions as you read.

3. Read. Read the section very carefully. As you read, try to understand the major concepts, and try to answer your previous questions. Highlight only the most important terms and concepts. Pay close attention to those terms that are already in boldface italic, since the authors have already considered them important.

4. Revise. After reading the section, revise your questions to more accurately reflect the section's contents. These revised questions should be conceptual, bringing together several important details. It may be helpful to write these questions in the margin of your textbook.

5. Record. Underline or highlight the information in the textbook that answers your questions. You may want to write the answers in your own form of shorthand. This process will give you good material for preparing for examinations.

6. Review. Review the information by trying to answer your questions without looking at the textbook. Use the questions in your textbook as well. Review the material several times.

PREPARING FOR EXAMINATIONS

It is important to prepare for examinations properly, so that you are not rushed and tired on examination day. Cramming at the last moment for an examination is no substitute for continued daily preparation and review. In fact, studies show that last-minute cramming and late-night studying are usually ineffective and counterproductive.

Complete your textbook reading and lecture-note revision ahead of time, and spend the last few days mastering the anatomical material, and not in trying to understand the basic concepts.

Good luck!

John P. Harley, Ph.D.

STUDENT'S STUDY MANUAL
to accompany *CAROLA, HARLEY, AND NOBACK*
Human Anatomy

Introduction to Anatomy

1

Prefixes

ana-	apart
cardio-	heart
cyto-	cell
extra-	outside of
hist-	fabric
homeo-	constant; equal
intra-	within
macro-	large
micro-	small
pariet-	wall
pector-	the breast
pelv-	pelvis; a basin
pleura-	the side
retro-	behind; back of
scop-	to look at
super-	above
ultra-	beyond

Suffixes

-graph	written
-logy	the study of
-stasis	standing; staying
-tomy	cutting

Read Chapter 1 in the textbook, focusing on the major objectives. As you progress through it, with your textbook open, answer and/or complete the following. When you complete this exercise, you will have a thorough outline of the chapter.

DEVELOPING YOUR OUTLINE

I. **Introduction** (p. 3)

    1 Why do we study the human body?

II. **What Is Anatomy?** (p. 3)

OBJECTIVE 1

    1 Explain the science of anatomy.

    2 The study of anatomy is divided into several areas. Describe each of the following areas of study:

    a regional anatomy

    b systemic anatomy

    c gross anatomy

    d microscopic anatomy

    e embryological anatomy

    f developmental anatomy

    g radiographic anatomy

  3 Give an example that illustrates the fact that function is determined by structure.

III. From Atom to Organism: Structural Levels of the Body (p. 7)

OBJECTIVE 2

  1 List, in order, the seven levels of body organization.

    _____, _____, _____, _____, _____,

    _____, _____ .

  A Atoms, Molecules, and Compounds (p. 7)
    1 Describe an atom.

    2 Define a compound and give two examples.
      a definition

      b examples

  B Cells (p. 7)
    1 What is a cell?

  C Tissues (p. 7)
    1 List the four major tissue types, give an example where each one is found in the body, and give at least one function for each tissue.

OBJECTIVE 3

Tissue type	Location	Function
_____	_____	_____
_____	_____	_____
_____	_____	_____
_____	_____	_____

D Organs (p. 9)

    1 List four organs found within the human body and give one function for each.

Organ	Function
_____	_____
_____	_____
_____	_____
_____	_____

E Systems (p. 9)

    1 Describe a typical system found within the human body.

IV. Body Systems (p. 9)

OBJECTIVE 4

    A Integumentary System (Figure 1.4)

        1 What are the different parts of the integumentary system?

        2 Describe the major and minor functions of the integumentary system.

    B Skeletal System (Figure 1.5)

        1 What are the major parts of the skeletal system?

        2 Describe five functions of the skeletal system.

            a

            b

            c

            d

            e

    C Muscular System (Figure 1.6)

        1 What are the three different muscle types found in the body?

            a

b

c

2 What are four specific functions of the muscular system?

　a

　b

　c

　d

D Nervous System (Figure 1.7)
　1 What are the different parts of the nervous system?

E Endocrine System (Figure 1.8)
　1 List the major components that comprise this system.

　2 Describe the major function of the endocrine system and how hormones
　　act to maintain this function.

F Cardiovascular System (Figure 1.9)
　1 List the components and their roles in this system.

　2 Describe some functions of blood.

G Lymphatic System (Figure 1.10)
　1 What anatomical parts comprise this system?

　2 Describe four functions of this system.
　　a

　　b

　　c

　　d

H Respiratory System (Figure 1.11)
    1 List the major components of this system.

    2 Describe the main function of this system.

I Digestive System (Figure 1.12)
    1 What are the anatomical parts of this system?

    2 Describe the main function of this system.

J Urinary System (Figure 1.13)
    1 What are the anatomical parts of this system?

    2 Describe the main functions of this system and their implications.

K Reproductive Systems (Figure 1.14)
    1 What are the anatomical parts of this system in the male? In the female?

    2 Describe the major functions of these systems.

V. Anatomical Terminology (p. 16)

OBJECTIVE 5

  A Anatomical Position (p. 16)
    1 What is meant by the "anatomical position" of the body?

    2 What is meant by the term "relative" with respect to directional terms?

    3 Indicate the anatomical position of each of the following terms:

OBJECTIVE 6

       superior

       inferior

       anterior

       posterior

       medial

       lateral

       proximal

distal

superficial

deep

external

internal

peripheral

plantar

dorsal

palmar

parietal

visceral

B Body Regions (p. 18)
    1 What are the main divisions of the body?

    2 What specific anatomical regions are found within each of the following?
      a upper abdomen

      b middle abdomen

      c lower abdomen

C Body Planes (p. 20)
    1 Describe each of the following planes:
      a midsagittal

      b sagittal

      c frontal (coronal)

      d transverse

    2 Describe the following sections:
      a midsagittal

    b frontal

    c transverse

D Body Cavities (p. 22)
    1 What are the two main body cavities?
      a

      b

    2 What are the two main parts of the ventral cavity?
      a

      b

    3 Describe the various parts of the thoracic cavity.

    4 What organs are found within the abdominopelvic cavity?

E Body Membranes (p. 23)
    1 What are membranes?

    2 List the three main types of membranes.
      a

      b

      c

VI. New Ways of Exploring the Body (p. 23)

OBJECTIVE 7

    1 What are two instruments that can be used to look inside the body?
      a

      b

A CAT (Computer-Assisted Tomography) and PET (Positron-Emission Tomography) Scanning (p. 23)
    1 What is the difference between a CAT scan and a PET scan?

B Dynamic Spatial Reconstructor (DSR) (p. 26)
    1 What does a DSR produce?

C Magnetic Resonance Imaging (MRI) (p. 26)
    1 What is another term for MRI?

D Digital Subtraction Angiography (DSA) (p. 26)
    1 What is DSA used for?

E Ultrasound (Sonography) (p. 27)
    1 Describe how ultrasound works.

F Thermography (p. 27)
    1 What disorders can be detected with thermography?

These are terms you should know before proceeding to the post-test.   **MAJOR TERMS**

anatomy *(p. 3)*

physiology *(p. 3)*

gross anatomy *(p. 3*

microscopic anatomy *(p. 3)*

atoms *(p. 7)*

molecule *(p. 7)*

compound *(p. 7)*

cells *(p. 7)*

tissues *(p. 7)*

organ *(p. 9)*

system *(p. 9)*

organism *(p. 9)*

anatomical position *(p. 16)*

superior *(p. 16)*

inferior *(p. 16)*

anterior (ventral) *(p. 16)*

posterior (dorsal) *(p. 16)*

medial *(p. 16)*

lateral *(p. 16)*

proximal *(p. 16)*

distal *(p. 16)*

superficial *(p. 16)*

deep *(p. 16)*

external *(p. 16)*

internal *(p. 16)*

peripheral *(p. 16)*

plantar *(p. 16)*

dorsal *(p. 16)*

palmar *(p. 16*

parietal *(p. 16)*

visceral *(p. 16)*

axial *(p. 18)*

appendicular *(p. 18)*

planes *(p. 20)*

midsagittal plane *(p. 20)*

sagittal plane *(p. 20)*

frontal (coronal) plane *(p. 20)*

transverse plane *(p. 20)*

midsagittal section *(p. 20)*

frontal section *(p. 20)*

transverse section *(p. 20)*

ventral cavity *(p. 22)*

dorsal cavity *(p. 22)*

thoracic cavity *(p. 22)*

abdominopelvic cavity *(p. 23)*

membranes *(p. 23)*

CAT scan *(p. 23)*

PET scan *(p. 24)*

dynamic spatial reconstructor *(p. 26)*

magnetic resonance imaging *(p. 26)*

digital subtraction angiography *(p. 26)*

ultrasound *(p. 27)*

thermography *(p. 27)*

MEDICAL LABORATORY TECHNOLOGIST AND TECHNICIAN

A **medical laboratory technologist (MLT)** administers and evaluates laboratory tests that play an important role in the detection, diagnosis, and treatment of disease. MLTs perform chemical, biological, microscopic, and microbiological tests. For example, they type blood, perform chemical tests to determine blood cholesterol levels, and conduct microscopic examinations of body fluids or tissue samples to detect the presence of microorganisms. Most medical technologists conduct tests related to the examination and treatment of patients. Others do research, develop laboratory techniques, teach, or perform administrative duties. A beginning job as a medical technologist requires four years of college training, including the completion of specialized program in medical technology. Most MLTs work in hospitals, clinics, or private laboratories.

Medical laboratory technicians generally have two years of college-level training. They perform a wide range of tests and laboratory procedures that require a high level of skill. However, unlike a medical technologist, the technician has not been trained to evaluate the many procedures and tests. Like technologists, medical laboratory technicians may work in several areas, specialize in one field, and work in hospitals, clinics, or private laboratories.

Based on the terminology contained within the chapter, label the following body cavities: **LABELING ACTIVITY**

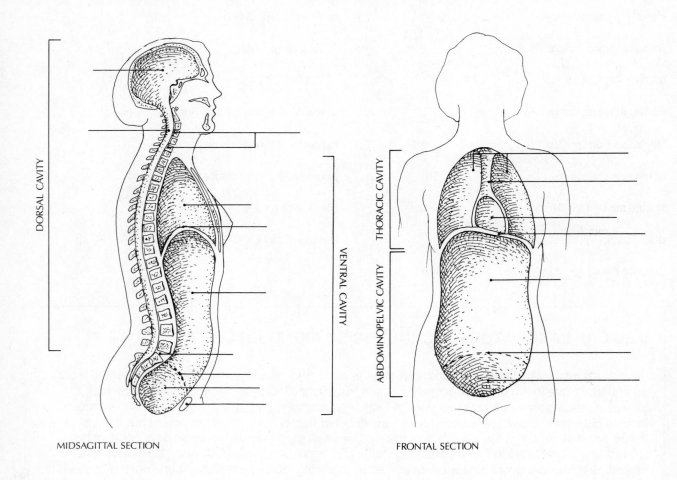

MIDSAGITTAL SECTION   FRONTAL SECTION

After you have completed the following activities, compare your answers with those given at the end of this manual. **POST-TEST**

1 In the anatomical position, the hand is _____ to the shoulder, whereas the shoulder is _____ to the hand.

    _____

    _____

2 Put the following terms in order from the lowest to the highest level of organization: atom, organism, system, cell, organ, tissue, molecule.

    _____ , _____ , _____ ,

    _____ , _____ , _____ ,

    _____ .

3 Matching.

   a ___ regional anatomy   [1] Requires the use of a microscope.   _____

   b ___ developmental anatomy   [2] The study of abnormal cells and tissues.   _____

   c ___ pathology   [3] The study of the head and neck.   _____

   d ___ microscopic anatomy   [4] The study of human growth (embryology).   _____

4 When structure and function are coordinated, our bodies are in a healthy state of   _____

   balance and stability called _____.

5 The ovaries are located within the: _____

   a pelvic cavity.

   b abdominal cavity.

   c thoracic cavity.

   d pleural cavity.

   e none of the above.

6 Which of the following is *not* part of the ventral cavity?   _____

   a abdominal

   b cranial

   c pericardial

   d thoracic

   e abdominopelvic

7 A _____ is a group of organs that works together to perform a   _____

   major body function.

8 Of the following, which is *not* one of the nine abdominal regions?   _____

   a umbilical

   b hypogastric

   c lumbar

   d hypochondriac

   e diaphragmatic

9 Which of the following body systems is mismatched?   _____

   a integumentary/protection

   b skeletal/movement

   c nervous/senses

   d endocrine/growth and development

   e cardiovascular/secretion

10 The pancreas, adrenal glands, and thymus gland are part of the _____ _____
system.

   a digestive

   b nervous

   c lymphatic

   d cardiovascular

   e endocrine

11 Matching.

   a ___ posterior   [1] Toward the front.   _____

   b ___ anterior   [2] Divides the body into identical halves.   _____

   c ___ midsagittal plane   [3] Nearer the trunk of the body.   _____

   d ___ lateral   [4] Toward the back.   _____

   e ___ proximal   [5] Away from a midsagittal plane.   _____

12 A _____ or coronal plane divides the body into anterior and   _____
posterior sections.

13 In anatomy, internal organs are commonly referred to as _____.   _____

14 The thoracic cavity is divided by membranes into what two cavities?

   a   _____

   b   _____

15 _____ are layers of epithelial and connective tissue that line the   _____
body cavities and cover or separate certain regions, structures, and organs.

16 Which of the following is *not* a main type of membrane?   _____

   a mucous

   b secretory

   c serous

   d synovial

   e cutaneous

17 Which of the following would you use to detect a blocked blood vessel?   _____

   a ultrasound

   b digital subtraction angiography

   c a CAT scan

   d a DSR scan

   e MRI

Surface Anatomy

2

Prefixes

acou-	hearing
acro-	extreme; peak
ante-	preceding
mamm-	breast
meta-	beyond; between
pelv-	basin
peri-	around
pleur-	rib
sub-	beneath

Suffixes

-ac	referring to cardiac
-ary	associated with
-ory	referring to auditory

Read Chapter 2 in the textbook, focusing on the major objectives. As you progress through it, with your textbook open, answer and/or complete the following. When you complete this exercise, you will have a complete outline of the chapter.

DEVELOPING YOUR OUTLINE

I. Introduction (p. 32)

1 How would you define surface anatomy?

2 Why is a study of surface anatomy useful to a nurse?

OBJECTIVE 1

3 Give the purpose of

OBJECTIVE 2

    a ascultation

    b inspection

    c palpation

    d percussion

OBJECTIVE 3

II. Head (p. 33)

  A Head (p. 33)

    1 What are the four parts of the cranial skull?

      a

      b

c

d

2 What are the four regions of the facial skull?

OBJECTIVE 4

a

b

c

d

3 Complete the following table on some of the surface structures of the head:

Structure	Description
parietal eminence	
nucha	
vertex	
symphysis menti	

III. Neck (p. 36)
A Neck (p. 36)

1 How would you define the neck?

2 Name and describe the two main triangles of the neck.

a

b

3 What is the clinical significance of the lymph nodes in the neck?

4 Complete the following table on some of the surface structures of the neck:

OBJECTIVE 5

Structure	Description
cricoid cartilage	_____
hyoid bone	_____
thyroid cartilage	_____
trachea	_____

IV. Trunk (p. 38)

OBJECTIVE 6

1 What are the different parts of the trunk?

A Thorax (p. 38)

1 How can the thorax be defined?

2 What is the apex heartbeat and where can it be heard best?

3 Complete the following table on some of the surface features of the thorax:

Structure	Description
anterior axillary fold	_____
areola	_____
costal margin	_____
sternum	_____

V. Back (p. 40)

A Back (p. 40)

1 How can the back be defined?

2 What is the triangle of auscultation?

3 Complete the following table on some of the surface features of the back:

Structure	Description
latissimus dorsi muscle	_____
posterior axillary fold	_____
scapula	_____
vertebra prominens	_____

VI. Abdomen and Pelvis (p. 42)

A Abdomen (p. 42)
   1 Where is the abdomen located?

   2 Complete the following table on some of the surface features of the abdomen:

Structure	Description
linea alba	_____
McBurney's point	_____
tendinous intersections	_____
mons pubis	_____

B Pelvis (p. 42)
   1 Where is the pelvis located?

   2 What is the pelvic girdle?

VII. Upper Extremity (p. 44)

OBJECTIVE 7

   1 What makes up the upper extremity?

A Pectoral Girdle and Axilla (p. 44)
   1 What bones make up the pectoral girdle?

   2 What is the axilla?

B Arm and Elbow (p. 44)

    1 Where is the arm located?

    2 Complete the following table on some of the surface features of the arm and elbow:

Structure	Description
brachial artery	_____
deltoid muscle	_____
elbow	_____
triceps brachi muscle	_____

    C The Forearm, Wrist, and Hand (p. 46)

        1 Define the

           a forearm

           b wrist

           c hand

        2 Complete the following table on some of the surface structures of the forearm, wrist, and hand:

Structure	Description
"anatomical snuffbox"	_____
knuckles	_____
palmar flexion creases	_____
thenar eminence	_____

VIII. Lower Extremity (p. 48)

    1 How would you define the lower extremity?

A Pelvic Girdle and Buttocks (p. 48)

    1 What three bones make up the pelvic girdle?

       a

       b

       c

    2 What are the buttocks?

    3 Complete the following table on some of the surface structures of the pelvic girdle and buttocks:

Structure	Description
femoral triangle	
gluteal cleft	
gluteal fold	
pubic tubercle	

B Thigh and Knee (p. 48)

    1 How would you define the thigh?

    2 Anatomically, what is the knee?

C Leg, Ankle, and Foot (p. 50)

    1 How would you define the leg?

    2 Complete the following table on some of the surface features of the leg, ankle, and foot:

Structure	Description
calcaneal (Achilles) tendon	
calcaneus	
dorsal venous arch	
tibia	

These are the terms you should know before proceeding to the post-test.   **MAJOR TERMS**

surface anatomy *(p. 32)*

percussing *(p. 32)*

palpating *(p. 32)*

head (cephalic region) *(p. 33)*

cranial skull *(p. 33)*

facial skull *(p. 33)*

neck *(p. 36)*

anterior triangle *(p. 36)*

posterior triangle *(p. 36)*

trunk *(p. 38)*

thorax *(p. 38)*

back *(p. 40)*

abdomen *(p. 42)*

pelvis *(p. 42)*

pelvic girdle *(p. 42)*

upper extremity *(p. 44)*

pectoral (shoulder) girdle *(p. 44)*

axilla *(p. 44)*

humerus *(p. 44)*

arm *(p. 44)*

elbow *(p. 44)*

forearm *(p. 46)*

wrist (carpus) *(p. 46)*

hand *(p. 46)*

lower extremity *(p. 48)*

buttocks *(p. 48)*

thigh *(p. 48)*

femur *(p. 48)*

knee *(p. 48)*

leg *(p. 50)*

ankle *(p. 50)*

foot *(p. 50)*

Based on the terminology contained within this chapter, label the following figures: **LABELING ACTIVITY**

1   The major anatomical regions of the cranial and facial skull.

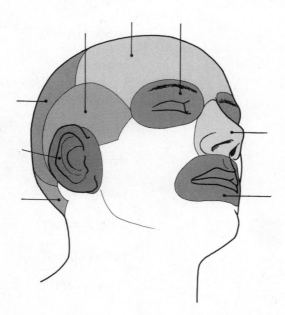

2   The head.

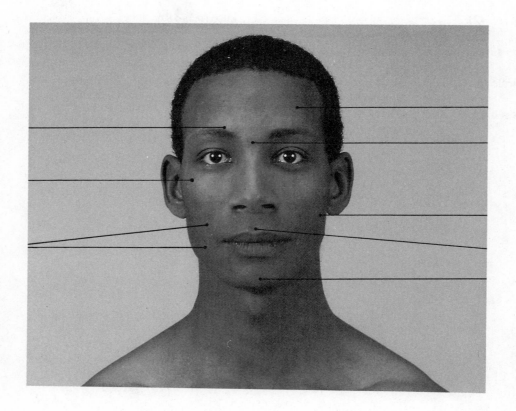

3   The neck.

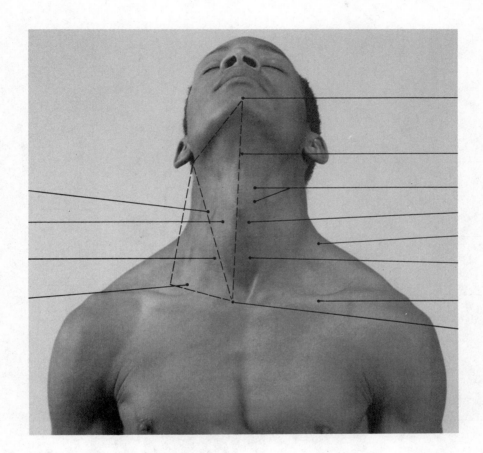

4 The thorax.

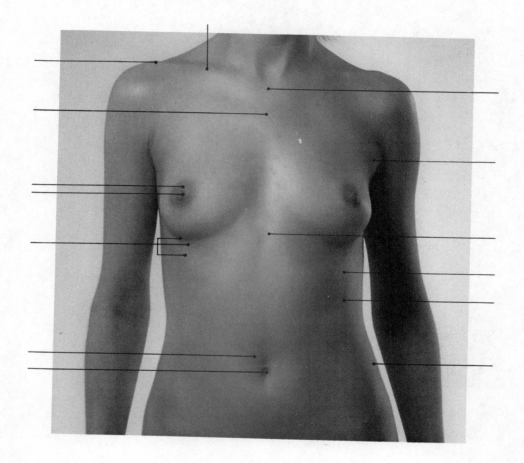

5   The back.

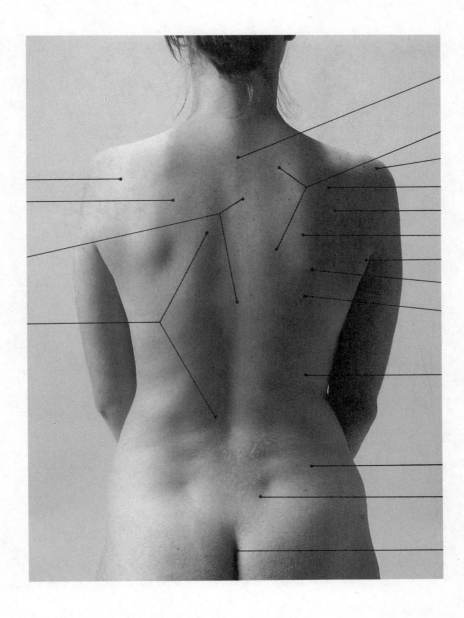

6   The abdomen.

7   The shoulder, axilla, arm, and elbow.

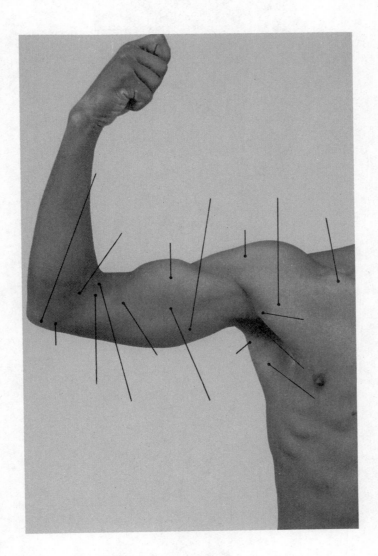

8   The forearm, wrist, and hand.

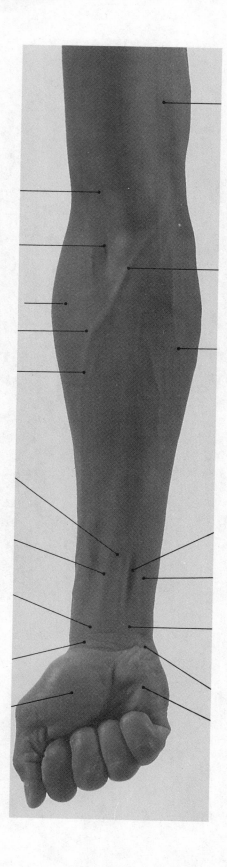

9   The buttocks and thigh.

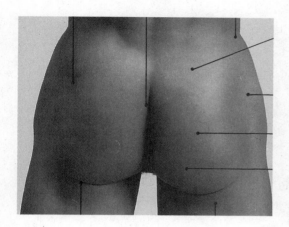

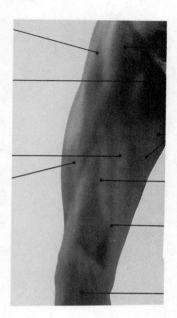

10   The leg and ankle.

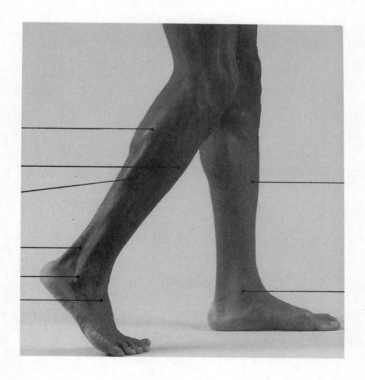

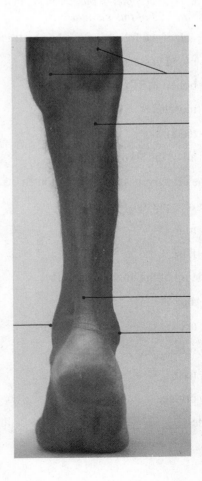

11   The foot.

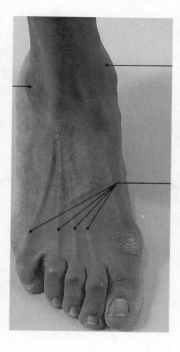

After you have completed the following activities, compare your answers with those given at the end of this manual.

POST-TEST

1   Which of the following can be identified from the surface of the body?

   a   blood vessels

   b   bones

   c   muscles

   d   tendons

   e   all of the above (**a–d**)

2   The effectiveness of observation and palpation may be altered by a large amount of fat on a person.

   a   true

   b   false

3   The tip of the nose is called the:

   a   ala.

   b   apex.

   c   root.

   d   nasi.

   e   bridge.

4   The _____ region of the skull contains the ears.

5 The skin of the forehead is part of the scalp.   _____

   a true

   b false

6 Eyeglasses are usually worn on the _____ of the nose.   _____

7 The inferior border of the mandible, the median line of the neck, and the   _____
   sternocleidomastoid muscle form the:

   a submandibular triangle.

   b superior triangle.

   c anterior triangle.

   d posterior triangle.

   e neck triangle.

8 The Adam's apple is anatomically the _____ cartilage.   _____

9 By observation, one can see several bony landmarks on the abdomen.   _____

   a true

   b false

10 The right costal margin of the rib cage is located over the liver.   _____

   a true

   b false

11 McBurney's point is an ideal entry point for an appendectomy.   _____

   a true

   b false

12 The _____ rib is located at the level of the sternal angle.   _____

13 The _____ fossa is located between the forearm and arm.   _____

14 The _____ _____ extends from the xiphoid process to the   _____
   symphysis pubis.   _____

15 The "anatomical snuffbox" is a depression located on the posterior surface of the   _____
   wrist when the thumb is extended.

   a true

   b false

16 The thenar eminence is located on the: _____

   a ulna.

   b calcaneus.

   c scapula.

   d humerus.

   e hand.

17 One takes a radial pulse on the arm near the wrist.   _____

    a true

    b false

18 Matching.

a ___ cricoid cartilage	[1] auricular	_____
b ___ costal margin	[2] thorax	_____
c ___ tragus	[3] hand	_____
d ___ philtrum	[4] oral	_____
e ___ thenar eminence	[5] neck	_____

19 Matching.

a ___ buttock	[1] IV therapy	_____
b ___ radial artery	[2] IM injection	_____
c ___ linea alba	[3] abdomen	_____
d ___ cubital vein	[4] pulse rate	_____
e ___ brachial artery	[5] pressure point	_____

20 The _____ point is an important point when palpating the appendix.   _____

21 The depression on the upper lip is called the _____.   _____

22 The region called the _____ contains the tibia bone.   _____

23 Anatomical features can be felt beneath the skin via _____.   _____

24 The hamstring muscles are located in the thigh.   _____

    a true

    b false

25 All of the ribs can be palpated.   _____

    a true

    b false

Cells: The Basic Units of Life

3

Prefixes

bi-	two; double
crist-	crest; fold
cyt-	cell
endo-	within
inter-	between
intra-	within
lys-	break apart; cut
mit-	thread
nucl-	pertaining to the nucleus
phago-	eat
pino-	drink
reticul-	network
syn-	together

Suffixes

-plasm	fluid
-some	body

Read Chapter 3 in the textbook, focusing on the major objectives. As you progress through it, with your textbook open, answer and/or complete the following. When you complete this exercise, you will have a thorough outline of the chapter.

DEVELOPING YOUR OUTLINE

I. Introduction (p. 55)

1 Why are cells the basic units of life?

II. What Are Cells? (p. 55)

OBJECTIVE 1

1 List the four basic parts of a cell.

a

b

c

d

2 Complete the following table of the cell's various structures, descriptions, and functions. Insert your answer in the spaces provided for each entry.

Cell structure	Description	Function
Plasma membrane	_____	_____
_____	Semifluid within plasma membrane; contains organelles.	_____
_____	_____	Site of protein synthesis.
Lysosomes	_____	_____
Microtubules	_____	_____
_____	_____	Provide for structural support and movement.

OBJECTIVE 2

III. Cell Membranes (p. 57)

A Structure of Cell Membranes (p. 57)

1 Using the following terms, draw and label a representation of a portion of a cell membrane (fluid-mosaic model) in the space provided.

surface carbohydrate
glycoprotein
phospholipid bilayer
phosphate heads
hydrophobic end
hydrophilic end
glycocalyx

B Functions of Cell Membranes (p. 57)

1 What are five functions of cell membranes?

a

b

c

d

e

2 Describe the selective permeability of a cell membrane.

3 What is the function of microvilli?

IV. Movement Across Membranes (p. 58)

1 Describe passive transport of materials across a cell membrane.

2 What are the four major types of passive transport processes that occur across a cell membrane?

a

b

c

d

3 Describe each of the following types of active movement across cell membranes:

a active transport

b exocytosis

c endocytosis

d pinocytosis

e phagocytosis

f receptor-mediated endocytosis

V. Cytoplasm (p. 61)

1 What are two parts of the cytoplasm?

a

b

VI. Organelles (p. 61)

    1 Describe an organelle.

    A Endoplasmic Reticulum (p. 61)

        1 Describe this organelle.

        2 What is the structure of the ER?

        3 What is the major difference between smooth and rough ER?

    B Ribosomes (p. 61)

        1 What is the function of this organelle?

        2 What are polyribosomes?

    C Golgi Apparatus (p. 63)

        1 What is the function of this organelle?

        2 Where are Golgi apparati most abundant?

        3 Describe the structure of the Golgi apparatus.

    D Lysosomes and Microbodies (p. 63)

        1 What is the function of a lysosome?

        2 What are the digestive enzymes of lysosomes called?

        3 What is one type of microbody present in most animal cells?

        4 How does the above microbody function?

    E Mitochondria (p. 64)

        1 Describe the major function of this organelle.

2 In the space below, sketch and label a mitochondrion.

F Microtubules, Intermediate Filaments, and Microfilaments (p. 64)
   1 Describe the cytoskeleton of a cell.

   2 Describe four roles microtubules play within the cell.
       a

       b

       c

       d

   3 Describe an intermediate filament.

   4 Describe a microfilament.

G Centrioles (p. 65)
   1 What is the relationship between centrioles and chromosomes?

   2 Describe or sketch the structure of this organelle.

H Cilia and Flagella (p. 65)
   1 Differentiate between cilia and flagella.

2 In the human body, where are ciliated cells located?

3 In the human body, where are flagellated cells located?

I Cytoplasmic Inclusions (p. 66)
    1 Describe several cytoplasmic inclusions.

VII. The Nucleus (p. 67)
    1 List the two important roles of the cell nucleus.

        a

        b

A Nuclear Envelope (p. 68)
    1 What is the function of the pores in the nuclear envelope?

    2 Describe the structure of the nuclear envelope.

B Nucleolus (p. 67)
    1 Describe the structure of the nucleolus.

C Chromosomes (p. 68)   **OBJECTIVE 3**
    1 What is chromatin?

    2 What is a chromosome?

D The Structure of Nucleic Acids (p. 69)   **OBJECTIVE 4**
    1 What is a gene?

    2 What is a chromosome?

VIII. The Cell Cycle (p. 69)   **OBJECTIVE 5**
    1 What is the cell cycle?

A Interphase (p. 71)

    1 What are the three distinct stages of interphase?

      a

      b

      c

B Mitosis (p. 71)

    1 What two things are accomplished by mitosis?

      a

      b

    2 Describe and sketch each of the following events in mitosis:

      a prophase

      b metaphase

      c anaphase

      d telophase

C Cytokinesis (p. 74)

    1 Describe what happens during cytoplasmic division.

IX. Meiosis (p. 75)

    1 Describe meiosis.

A Comparison of Mitosis and Meiosis (p. 75)

    1 Contrast mitosis with meiosis.

X. Development of Primary Germ Layers (p. 75)

    1 Define differentiation.

2 What are the three germ layers?

a

b

c

XI. Cells In Transition: Aging (p. 77)

OBJECTIVE 6

1 What is the difference between necrobiosis and necrosis?

2 What is gerontology?

A Hypotheses of Aging (p. 77)

1 Describe six hypotheses of aging.

a

b

c

d

e

f

XII. Cells Out of Control: Cancer (p. 78)

OBJECTIVE 7

1 How can cancer be defined?

A Neoplasms (p. 78)

1 Define a neoplasm.

2 What is the difference between a benign and malignant neoplasm?

3 Describe each of the following:

a carcinoma

b sarcoma

    c mixed-tissue neoplasm

    d leukemia

B How Cancer Spreads (p. 78)
    1 What does the word *metastasis* mean?

    2 What is the difference between a primary and secondary neoplasm?

C Causes of Cancer (p. 79)
    1 What is a carcinogen?

    2 What are some causes of cancer?

    3 What is an onogene?

D Treatment of Cancer (p. 80)
    1 What are some ways that cancer can be treated?

These are the terms you should know before proceeding to the post-test.   **MAJOR TERMS**

protoplasm *(p. 55)*	particulate phase *(p. 61)*
nucleus *(p. 55)*	aqueous phase *(p. 61)*
nucleoplasm *(p. 55)*	organelles *(p. 61)*
cytoplasm *(p. 55)*	endoplasmic reticulum *(p. 61)*
plasma membrane *(p. 57)*	ribosomes *(p. 61)*
fluid-mosaic model *(p. 57)*	Golgi (apparatus) complex *(p. 63)*
phospholipids *(p. 57)*	lysosomes *(p. 63)*
bimolecular layer (bilayer) *(p. 57)*	microbodies *(p. 63)*
glycocalyx (cell coat) *(p. 57)*	peroxisomes *(p. 64)*
selective permeability *(p. 57)*	mitochondria *(p. 64)*
passive transport *(p. 58)*	cytoskeleton *(p. 64)*
concentration gradient *(p. 58)*	microtubules *(p. 64)*

intermediate filaments *(p. 65)*

microfilaments *(p. 65)*

centrioles *(p. 65)*

cilia *(p. 65)*

flagella *(p. 65)*

cytoplasmic inclusions *(p. 66)*

nuclear envelope *(p. 67)*

nucleolus *(p. 67)*

chromosomes *(p. 68)*

cell cycle *(p. 69)*

replication *(p. 69)*

nucleotides *(p. 69)*

nitrogenous base *(p. 69)*

gene *(p. 69)*

interphase *(p. 71)*

G1 *(p. 71)*

S *(p. 71)*

G2 *(p. 71)*

mitosis *(p. 71)*

prophase *(p. 72)*

chromatids *(p. 72)*

metaphase *(p. 73)*

anaphase *(p. 73)*

telophase *(p. 74)*

cytokinesis *(p. 74)*

meiosis *(p. 75)*

diploid *(p. 75)*

haploid *(p. 75)*

homologous pair (tetrad) *(p. 75)*

crossing-over *(p. 75)*

differentiation *(p. 75)*

primary germ layers *(p. 75)*

ectoderm *(p. 76)*

endoderm *(p. 76)*

embryonic disk *(p. 77)*

mesoderm *(p. 77)*

gerontology *(p. 77)*

necrobiosis *(p. 77)*

necrosis *(p. 77)*

free-radical damage *(p. 77)*

cell senescence *(p. 77)*

mutations *(p. 77)*

glycosylation *(p. 77)*

neoplasm *(p. 78)*

benign neoplasm *(p. 78)*

malignant neoplasm *(p. 78)*

carcinoma *(p. 78)*

sarcoma *(p. 78)*

mixed-tissue neoplasm *(p. 78)*

leukemias *(p. 78)*

metastasis *(p. 78)*

primary neoplasm *(p. 78)*

secondary neoplasm *(p. 78)*

dysplasia *(p. 78)*

carcinoma in situ *(p. 78)*   carcinogens *(p. 79)*

malignant carcinoma *(p. 79)*   oncogene *(p. 80)*

MEDICAL RECORD ADMINISTRATOR, MEDICAL RECORD TECHNICIAN, AND CYTOTECHNOLOGIST

Medical record administrators manage record departments and develop systems for documenting, storing, and retrieving medical information. They supervise and train medical record technicians and clerks, compile medical statistics, and help evaluate patient care and research studies. Most medical record administrators work in hospitals or coordinate the medical record departments of several small hospitals. Others work in clinics, nursing homes, state and local public health departments, medical research centers, and health insurance companies. Some work for firms that develop and print health insurance and medical forms, others work for companies that manufacture equipment to record and process medical data. Still others are consultants to small health care facilities. Medical record administrators can advance to positions as department heads in large hospitals or to higher-level positions in hospital administration. Often, people in these positions have gained experience in smaller health care facilities. Many medical record administrators teach in the expanding two- and four-year college programs for medical record personnel.

Most medical record administrators have a bachelor's degree in medical record administration. Those who have a bachelor's degree in another field and the required courses in the liberal arts and biological sciences may complete a one-year certificate program. Students interested in this field should take courses in health, business, mathematics, and biology.

Medical record technicians serve as technical assistants to the registered record administrator and carry out many technical activities within a medical record department. The duties of medical record technicians vary with the size of the institution. In a small institution, the accredited medical record technician may have full responsibility for the operation of this department—compiling, analyzing, and preparing health information needed by the patient, the health care facility, third parties, and the public. In a large institution, the technician may specialize in a particular phase of work performed in the medical record department. Most medical record technician positions require a high school diploma or the equivalent.

Cytotechnologists are trained medical laboratory technologists who work with pathologists to detect changes in body cells that may be important n the early diagnosis of cancer. This is done primarily by examining slide preparations of body cells with a microscope. The cytotechnologist looks for abnormalities in cell structure that might indicate either benign or malignant conditions. The clinical education is a minimum of 12 months.

Based on the terminology contained within the chapter, label the following cell: **LABELING ACTIVITY**

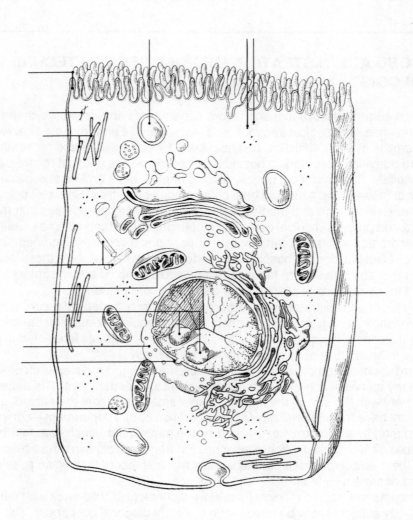

After you have completed the following activities, compare your answers with those given at the end of this manual.   **POST-TEST**

1 The cell is the basic structural unit of the body   _____

    a true

    b false

2 Which of the following transport processes requires ATP energy?   _____

    a diffusion

    b facilitated diffusion

    c osmosis

    d filtration

    e active transport

3 Which of the following describes the fluid mosaic model of the plasma membrane? _____

   a double lipid layer that loosely surrounds globular proteins that extend beyond the membrane's surface

   b a mosaic of proteins and lipids

   c hydrophobic and hydrophilic protein, lipid, and carbohydrate

   d a and b

   e a, b and c

4 What are the four basic regions of a cell? _____

   a _____

   b _____

   c _____

   d

5 Ribosomes package proteins for secretion. _____

   a true

   b false

6 The Golgi apparatus packages proteins for cell secretion. _____

   a true

   b false

7 Matching.

   a ___ prophase   [1] Nuclear membrane forms around _____
                                 chromosome set.

   b ___ metaphase   [2] Chromatin forms chromosomes. _____

   c ___ anaphase   [3] Chromosomes separate. _____

   d ___ telophase   [4] The centrioles migrate and the _____
                                   microtubules attach to the chromatids.

8 The _____ provides for structural support and cell movement.

9 _____ store cellular material. _____

10 The _____ is the preassembly point for ribosomes. _____

11 The _____ houses the cellular organelles. _____

12 A metabolically active cell will: _____

   a contain a large number of mitochondria. _____

   b have nuclei.

   c contain no ER.

   d have many lysosomes.

   e all of the above.

13 Pairs of microtubules arranged in a circle around two microtubules would be found in: _____

    a microfilaments.

    b cilia.

    c microtubules.

    d a and **b**

    e a, **b,** and **c.**

14 That part of the cytoplasm that supports the organelles and inclusions is called the _____

_____.

Tissues of the Body

4

Prefixes

ab-	away from
acin-	grape; sac
adip-	fat
apo-	away from; off
chondro-	cartilage
crin-	to separate
desmo-	band; ligament
epi-	upon; over
histo-	web
holo-	entire
meso-	middle
micro-	small
neuro-	nerve
osseo-	bone
phag-	to eat
pseudo-	false

Suffixes

-blast	formative
-clast	broken
-cyte	hollow vessel
-glia	glue
-oma	tumor

Read Chapter 4 in the textbook, focusing on the major objectives. As you progress through it, with your textbook open, answer and/or complete the following. When you complete this exercise, you will have a thorough outline of the chapter.

DEVELOPING YOUR OUTLINE

I. Introduction (p. 85)

1 Why is it necessary to study histology in a human anatomy course?

2 Define a tissue.

OBJECTIVE 1

II. Epithelial Tissues (p. 85)

1 In general, epithelial tissues have two anatomical functions. What are they?

OBJECTIVE 2

a

b

2 How are epithelial tissues arranged?

A Functions of Epithelial Tissues (p. 85)

    1 List the six major functions of epithelial tissues.

        a

        b

        c

        d

        e

        f

B General Characteristics of Epithelial Tissues (p. 85)

    1 What are the three main junctional complexes that hold epithelial cells together? Describe each one briefly.

        a

        b

        c

    2 What are the principal parts of the basement membrane (basal lamina)?

    3 Since epithelial tissues do not contain blood vessels, how do oxygen and nutrients get to the cells?

    4 What are three surface specializations of epithelial tissues? What function does each specialization perform?

        a

        b

        c

    5 How do epithelial tissues regenerate themselves?

C General Classification of Epithelial Tissues (p. 87)

1 What three criteria are used to classify epithelial tissues?

a

b

c

2 List the five major types of epithelial tissues based on shape and arrangement.

a

b

c

d

e

3 Describe the structure and major function(s) of each of the following epithelial tissues:

a simple squamous

b simple cuboidal

c simple columnar

d stratified squamous

e stratified cuboidal

f stratified columnar

D Exocrine Glands (p. 93)

1 Define an exocrine gland.

2 What are the five types of multicellular simple exocrine glands?

a

b

c

d

e

3 What are the three types of multicellular compound glands?

a

b

c

III. Connective Tissues (p. 95)

A Fibers of Connective Tissues (p. 95)

    1 List and describe the three major types of fibers found in connective tissues.

      a

      b

      c

B Ground Substance of Connective Tissues (p. 96)

    1 What is the structure of connective tissue ground substance?

    2 What is its function?

    3 What are the major ingredients of ground substance?

C Cells of Connective Tissues (p. 96)

    1 List the four types of fixed connective tissue cells and give the structure
        and function for each type.

Type	Structure and function
a _____	
b _____	
c _____	
d _____	

    2 List the four types of wandering cells found in connective tissue and
        give the structure and function of each type.

Type	Structure and function
a _____	
b _____	
c _____	
d _____	

D Connective Tissues Proper (p. 97)

    1 List the major types of connective tissues and give the structure and function for each type.

Type	Structure and function
a	_____
b	_____
c	_____
d	_____
e	_____
f	_____
g	_____
h	_____
i	_____

    2 What are two types of dense connective tissue?

      a

      b

E Cartilage (p. 101)

    1 What are cartilage cells called?

    2 Describe several types of cartilage and give a function for each type.

F Bone As Connective Tissue (p. 101)

    1 How does bone differ from cartilage?

G Blood As Connective Tissue (p. 101)

    1 Why is blood a type of connective tissue?

2 What are the three components of blood that all connective tissues have?

a

b

c

IV. Muscle Tissue: An Introduction (p. 101)

OBJECTIVE 6

1 Name the three types of muscle tissue.

a

b

c

V. Nervous Tissue: An Introduction (p. 101)

1 Name three cell types found n nervous tissue.

a

b

c

VI. Membranes (p. 101)

A Mucous Membranes (p. 101)

1 What are some functions of mucus?

2 Where are mucous membranes found in the human body?

B Serous Membranes (p. 101)

1 Describe a serous membrane.

2 Where in the human body are serous membranes found?

3 What are mesenteries?

4 Name the three serous membranes that are found within the human body.

a

b

c

C Synovial Membranes (p. 103)

1 Where are synovial membranes found in the human body?

2 Give a brief description of a synovial membrane.

VII. When Things Go Wrong (p. 104)

1 List three diseases that result from faulty collagen.

a

b

c

These are terms you should know before proceeding to the post-test.   **MAJOR TERMS**

epithelial tissue (epithelia) *(p. 85)*   pseudostratified columnar epithelium *(p. 92)*

junctional complexes *(p. 85)*   transitional epithelium *(p. 92)*

basement membrane *(p. 85)*   exocrine glands *(p. 93)*

microvilli *(p. 86)*   unicellular gland *(p. 93)*

brush border *(p. 86)*   goblet cell *(p. 93)*

cilia *(p. 86)*   multicellular gland *(p. 93)*

simple squamous epithelium *(p. 88)*   simple gland *(p. 93)*

simple cuboidal epithelium *(p. 88)*   compound gland *(p. 93)*

simple columnar epithelium *(p. 89)*   tubular gland *(p. 93)*

stratified squamous epithelium *(p. 90)*   alveolar (acinar) gland *(p. 93)*

stratified cuboidal epithelium *(p. 91)*   tubuloalveolar gland *(p. 93)*

stratified columnar epithelium *(p. 91)*   mucous gland *(p. 93)* serous gland *(p. 93)*

mixed gland *(p. 93)*

merocrine gland *(p. 93)*

holocrine gland *(p. 93)*

apocrine gland *(p. 93)*

matrix *(p. 95)*

collagenous fibers *(p. 95)*

collagen *(p. 95)*

reticular fibers *(p. 96)*

elastic fibers *(p. 96)*

fixed cells *(p. 96)*

fibroblasts *(p. 96)*

adipose (fat) cells *(p. 96)*

macrophage cells *(p. 96)*

reticular cells *(p. 96)*

wandering cells *(p. 96)*

leukocytes (white blood cells) *(p. 96)*

plasma cells *(p. 96)*

mast cells *(p. 97)*

macrophages *(p. 97)*

reticuloendothelial system *(p. 97)*

cartilage *(p. 101)*

chondrocytes *(p. 101)*

lacunae *(p. 101)*

skeletal muscle *(p. 101)*

smooth muscle *(p. 101)*

cardiac muscle *(p. 101)*

nervous tissue *(p. 101)*

membranes *(p. 101)*

mucous membranes *(p. 101)*

serous membranes *(p. 101)*

peritoneum *(p. 101)*

pericardium *(p. 101)*

pleura *(p. 103)*

mesenteries *(p. 103)*

synovial membrane *(p. 103)*

collagen *(p. 104)*

scurvy *(p. 104)*

rheumatoid arthritis *(p. 104)*

arteriosclerosis *(p. 104)*

systemic lupus erythematosus *(p. 104)*

Marfan's syndrome *(p. 104)*

biopsy *(p. 104)*

allograft *(p. 104)*

heterograft *(p. 105)*

autograft *(p. 105)*

syngeneic graft *(p. 105)*

Based on the terminology contained within the chapter, label the following figures: **LABELING ACTIVITY**

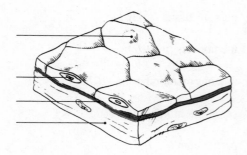

SIMPLE SQUAMOUS EPITHELIUM

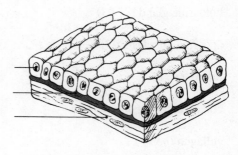

SIMPLE CUBOIDAL EPITHELIUM

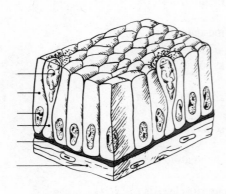

SIMPLE COLUMNAR EPITHELIUM

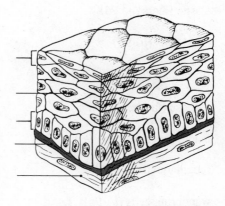

STRATIFIED SQUAMOUS EPITHELIUM

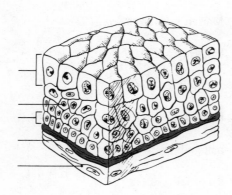

STRATIFIED CUBOIDAL EPITHELIUM

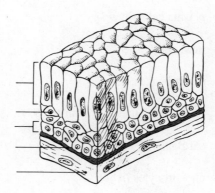

STRATIFIED COLUMNAR EPITHELIUM

To complete the following tables, place an X in the spaces where the answer is correct and leave blank the spaces where the answer is incorrect.   **POST-TEST**

1 Match the morphological categories of epithelial tissues with their general characteristics by placing an X in the appropriate spaces.

	Simple	Stratified	Pseudostratified	Transitional
More than one layer of cells	_____	_____	_____	_____
All cells in contact with basement membrane	_____	_____	_____	_____
Can be keratinized	_____	_____	_____	_____
Goblet cells may be present	_____	_____	_____	_____
Mesothelium	_____	_____	_____	_____

2 Which epithelial types can be found lining the following organs or structures in the human body?

	Simple	Stratified	Pseudostratified	Transitional
Peritoneum	_____	_____	_____	_____
Urinary bladder	_____	_____	_____	_____
Esophagus	_____	_____	_____	_____
Bronchi	_____	_____	_____	_____
Stomach	_____	_____	_____	_____

3 Which of the following is *not* a specialized function performed by epithelial tissues?   _____

a adsorption

b excretion

c secretion

d absorption

e protection

4 The three main types of junctional complexes are:

   a _____

   b _____

   c _____

_____

_____

_____

5 A single layer of epithelium that lines the inside of the heart is:

   a mesothelium.

   b endothelium.

   c pseudostratified epithelium.

   d columnar epithelium.

   e none of the above.

_____

6 Rounded, saclike exocrine glands are called:

   a tubular.

   b simple.

   c acinar.

   d compound.

   e a, b, and c

_____

7 Connective tissue fibers can be classified as:

   a reticular.

   b elastic.

   c collagenous.

   d a and b

   e a, b, and c

_____

8 Goblet cells secrete _____.

_____

9 A tissue specialized for fat storage is _____.

_____

10 Epithelia composed of more than one layer of cells are stratified epithelia.

   a true

   b false

_____

11 The goblet cell is actually a unicellular exocrine gland.

   a true

   b false

_____

12 Reticular fibers are formed from the fibrous protein elastin.

   a true

   b false

_____

13 Epithelial cells divide at a very slow rate.

   a true

   b false

14 Matching.

    a ___ fibroblast   [1] Gives skin color.   _____

    b ___ elastin fibers   [2] Stores fat.   _____

    c ___ adipose cell   [3] Produces collagen.   _____

    d ___ melanocyte   [4] Found in connective tissue. _____

15 Matching.

    a ___ secretion   [1] The release of molecules into   _____
                                          extracellular fluid.

    b ___ excretion   [2] Engulfs microorganisms.   _____

    c ___ microvilli   [3] Foldings of the plasma   _____
                                          membrane.

    d ___ mast cell   [4] Movement to external   _____
                                          environment.

16 In general, where in the human body are mucous membranes located?   _____

_____

17 Which of the following is *not* a type of cartilage?   _____

    a hyaline

    b elastic

    c fibrous

    d compact

    e **a** and **b**

18 Chondrocytes are found in: _____

    a ligaments.

    b epithelial tissues.

    c adipose tissue.

    d cartilage.

    e tendons.

19 Which of the following is *not* an area where simple squamous epithelium can be _____
    found?

    a lining the urinary bladder

    b lining the abdominal cavity

    c lining air sacs in the lungs

    d lining blood vessels

    e covering the skin

20 Collagenous connective tissue fibers contain the protein elastin.   _____

    a true

    b false

21 All connective tissues are initially derived from mesenchyme.   _____

    a true

    b false

22 Most cartilage is avascular.   _____

    a true

    b false

23 The interstitial matrix of hyaline cartilage contains: _____

    a proteoglycans.

    b keratin sulfate.

    c collagen fibers.

    d nerves.

    e **a** and **c.**

24 The main cell type found in connective tissue is the: _____

    a fibroblast.

    b goblet cell.

    c mast cell.

    d lymphocyte.

    e collagen cell.

25 Serous membranes include the: _____

    a pleura.

    b peritoneum.

    c pericardium.

    d **a** and **b.**

    e **a, b,** and **c.**

The Integumentary System

5

Prefixes

a-	root; without
alb-	white
corn-	horn
derm-	skin
epi-	on; upon
fasci-	band
follic-	small bag
hypo-	under
kerati-	horn
luc-	light
melan-	black; dark
papill-	nipple
por-	channel
seb-	grease
strat-	to spread, layer

Suffixes

-blast	germ
-cyte	cell
-phil	loving; dear

Read Chapter 5 in the textbook, focusing on the major objectives. As you progress through it, with your textbook open, answer and/or complete the following. When you complete this exercise, you will have a thorough outline of the chapter.

DEVELOPING YOUR OUTLINE

I. Introduction (p. 111)

1 What parts make up the integumentary system?

Skin, Hair, sebacious and sweat glands, nails

2 What are the five general functions of the integumentary system?

a

b

c

d

e

II. Skin (p. 111)

1 What are the two main parts of skin?

a epidermis

b dermis

A Epidermis (p. 111)

1 Describe the stratum corneum layer of the epidermis. multicellular, dandruff, dead skin

2 What is the function of keratin? Which layer of epidermis contains eleidin? makes surfaces tough, able to resist tension

3 What is unique about the stratum granulosum layer of the skin? waterproofing

4 What is the shape of the cells that make up the stratum spinosum layer of the skin? Star shaped cells

5 What is the shape of the cells that make up the stratum basale layer of the skin? spider shaped cells

6 Why is the stratum lucidum layer of the skin so named? Appears as a thin translucent Band

B Dermis (p. 111)

1 What types of fibers are found in the dermal layer of the skin? Nerve Fibers, blood vessels

2 What are the major cell types found in the dermal layer of the skin? Fibroblasts, macrophages, mast cells, scattered white Blood cells CT proper

3 What is another name for the papillary layer of the dermis? Areolar C.T.

4 What part of an animal's skin is commercially processed to make leather?

5 The arrangement and pattern of collagen fibers in the skin are known as

_____.

6 List some of the structures found within the hypodermis. fatty, below the skin

C Functions of the Skin (p. 114)

   1 How does the skin prevent entry of microorganisms and loss of body fluids?

   2 How does the skin help regulate body temperature?

       From Sweat glands

   3 List the ways heat can be lost from the body.

   4 What waste materials are excreted via perspiration?

   5 What vitamin does the skin synthesize?

   6 What are the special senses that the human skin has receptors for in order to help the body maintain homeostasis?

D Color of Skin (p. 115)

                                                               `OBJECTIVE 3`

   1 List the three factors that determine skin color.

     a *Melanin*

     b *Carotene*

     c *hemoglobin*

   2 What is the function of melanin? *imparts color to the skin and hair*

E How a Wound Heals (p. 115)

   1 What does each of the following contribute to the healing of a wound?

     a platelets — *helps initiate clotting*

     b neutrophils — *destroys bacteria*

     c fibroblasts — *secretes the fibers and ground substance of C.T. proper*

     d collagen — *replaces granulation tissue with scar tissue*

     e fibrinogen

     f monocytes *develops into macrophage which dispose of dead tissue cell*

2 Dead white blood cells and tissue debris in a wound form what type of fluid?

F Developmental Anatomy of the Skin (p. 116)
   1 How does skin develop?

III. Glands of the Skin (p.117)

   1 Name the two glands that are found in the skin.

   a

   b

A Sudoriferous (Sweat) Glands (p. 117)
   1 List the two major types of sudoriferous glands and their products.

   a

   b

   2 What is cerumen?

B Sebaceous (Oil) Glands (p. 118)
   1 Describe sebum.

   2 Describe the structure of a sebaceous gland.

   3 What is a blackhead?

IV. Hair (p. 118)

A Functions of Hair (p. 118)
   1 Describe four human functions of hair.

   a

   b

   c

   d

B Structure of Hair (p. 118)

   1 From the inside out, what are the three layers of a strand of hair?

     a *Medulla - central core*

     b *Cortex - surrounds medulla*

     c *Cuticle - atermost*

   2 List the three sheaths that comprise a hair follicle.

     a *Internal Root Sheath*

     b *External " "*

     c *Connective Tissue Root Sheath*

   3 What determines the shape and texture of one's hair?

     vellus or terminal

   4 What are the two effects of the contraction of the arrector pili muscle?

     a *hair stands errect*

     b *skin gets goose pumps*

C Growth, Loss, and Replacement of Hair (p. 119)

   1 When does hair grow the fastest? The slowest?

     spring   *winter*

   2 Describe pattern baldness.

     genetic

D Developmental Anatomy of Hair (p. 120)

   1 Describe each of the following types of hair:

     a lanugo

     b vellus

     c coarse

     d terminal

OBJECTIVE 6

V. Nails (p. 120)

    1   Why do nails appear pink?

rich capillaries in the dermis

    2   Describe the function of the nail matrix.

    3   In humans, what are some functions of nails?

pick up small objects, scratch

    4   In the space below, draw and label the structure of a nail.

VI. Effects of Aging on the Integumentary System (p. 121)

    1   What happens to the skin when it ages?

VII. When Things Go Wrong (p. 121)

    A   Burns (p. 121

       1   How are burns classified?

       2   Explain the Lund-Browder method.

       3   Describe the "rules of nine."

       4   Complete the following table:

Type of burn	Amount of surface area affected	Major effects	Depth of tissue damage
first-degree	_____	_____	_____
second-degree	_____	_____	_____
third-degree	_____	_____	_____

B Some Common Skin Disorders (p. 123)

    1 Complete the following table on skin disorders:

Disease	Examples or symptom	Cause
acne vulgaris		
bedsores		
birthmarks		
moles		
psoriasis		
allergic responses		
warts		
skin cancer		
bruises		

These are the terms you should know before proceeding to the post-test.   **MAJOR TERMS**

integumentary system *(p. 111)*

skin *(p. 111)*

epidermis *(p. 111)*

stratum corneum *(p. 111)* ~~dandruff and flakes~~

stratum lucidum *(p. 111)* ~~clear layer flat dead keratinocytes~~

stratum granulosum *(p. 111)*

stratum spinosum *(p. 111)*

stratum basale *(p. 111)*

dermis *(p. 111)* ~~second layer of skin strong CT.~~

papillary layer *(p. 113)*

reticular layer *(p. 113)*

hypodermis (subcutaneous) layer *(p. 114)*

melanin *(p. 115)*

sudoriferous (sweat) gland *(p. 117)*

eccrine gland *(p. 117)*

apocrine (oderiferous) gland *(p. 117)*

sebaceous (oil) gland *(p. 118)*

sebum *(p. 118)*

blackhead *(p. 118)*

hair *(p. 118)*

shaft *(p. 118)*

root *(p. 118)*

medulla *(p. 118)*

cortex *(p. 118)*

cuticle *(p. 118)*

matrix *(p. 118)*

follicle *(p. 118)*

arrector pili muscle *(p. 119)*

ATHLETIC TRAINER

Athletic trainers are trained professionals who work to prevent and treat sports-related injuries. They work with physicians to care for athletes. The demand for athletic trainers has increased greatly as a result of the enormous growth of sports, from high school to professional levels. With the increase of participation in sports, there has been a rise in injuries. Athletic trainers supervise practices, give advice on the prevention of injuries, tape areas of the body that are susceptible to injury or stress, and attend to any injuries that may occur during sporting events. Trainers work for high schools, colleges, and professional athletic teams and must be certified by the Association for Athletic Trainers. Certification programs vary, but all are at the college level and include the study of anatomy and physiology in addition to first aid and rehabilitation procedures.

The Integumentary System

Based on the terminology contained within this chapter, label the following figure of the skin:

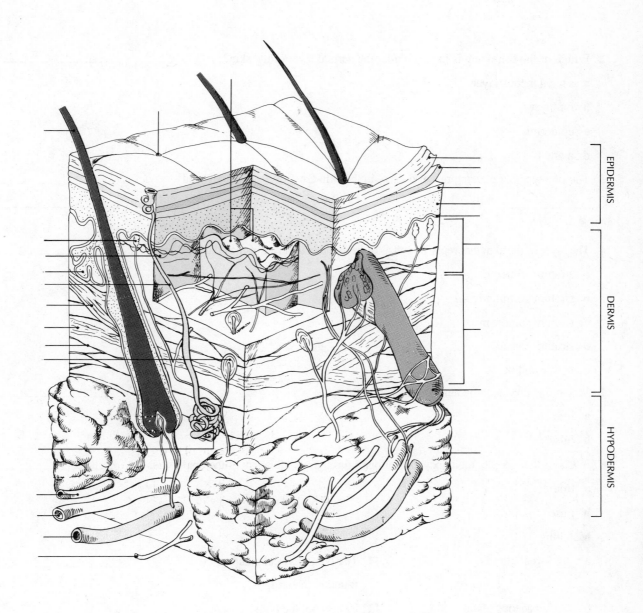

EPIDERMIS

DERMIS

HYPODERMIS

After you have completed the following activities, compare your answers with those given at the end of this manual.

1 Why is the skin considered an organ?

2 Which of the following is located under the dermal layer of the skin?   _____

   a subcutaneous layer

   b fat layer

   c epidermis

   d **a** and **b**

3 The two types of glands present present in the skin are _____ and   _____

   _____ .   _____

4 The layer of the skin that replaces cells lost from the outer layer of the epidermis is the: _____

   a stratum corneum.

   b stratum germinativum.

   c stratum spinosum.

   d stratum basale.

   e both **c** and **d.**

5 Melanin is produced by keratin cells.   _____

   a true

   b false

6 In Caucasian people, melanocytes are found throughout all layers of the epidermis.   _____

   a true

   b false

7 Matching.

   a ___ sweat gland   [1] The type of sweat gland   _____

                         found in the armpits.

   b ___ sebaceous gland   [2] Helps cool the body.   _____

   c ___ apocrine gland   [3] Secretes an oil that   _____

                         lubricates the skin.

   d ___ eccrine gland   [4] The most common type   _____

                         of sweat gland.

8 Which of the following is *not* a function of the skin?   _____

   a temperature regulation

   b protection

   c sensation

   d absorption and synthesis

   e storage

9 Put the following in the correct sequence: _____

   a The most serious to least serious burns.   _____

     i third-degree   _____

     ii second-degree

     iii first-degree

   b From the outside to the inside of a strand of hair.   _____

     i cortex   _____

     ii medulla   _____

     iii cuticle

   c The most dangerous to least dangerous tumor.   _____

     i basal-cell epithelioma   _____

     ii squamous-cell carcinoma   _____

     iii malignant melanoma

   d from superficial to deep layers of skin.   _____

     i hypodermis   _____

     ii epidermis   _____

     iii dermis

10 Which of the following is composed of tough keratin?   _____

   a melanocytes

   b hair follicles

   c nails

   d sweat glands

   e sebum

11 Ultraviolet rays from the sun are absorbed by: _____

   a the basal layer of the skin.

   b melanocytes.

   c melanin.

   d hair papillae.

   e all of the above.

12 During wound healing, collagen is produced by: _____

   a fibroblasts.

   b epithelial cells.

   c melanocytes.

   d monocytes.

   e collagen cells.

13 Sebaceous glands are commonly known as: _____

   a oil glands.

   b apocrine glands.

   c sweat glands.

   d a and b.

   e b and c.

14 The active, growing part of the nail is the _____ .   _____

15 The glands associated with hair follicles are called: _____

   a exocrine glands.

   b eccrine glands.

   c sebaceous glands.

   d endocrine glands.

   e sweat glands.

16 Blood vessels that give skin its color are located in the _____ layer.   _____

17 The scientific term for a bedsore is _____ .   _____

18 Warts (verrucae) are benign epithelial tumors caused by _____ .   _____

19 The appearance of fingerprints is determined by the contours of the dermal papillae.   _____

   a true

   b false

20 The muscle that causes goose bumps is the: _____

   a terminal muscle.

   b errector pili muscle.

   c arrector pili muscle.

   d a and b.

   e a, b, and c.

21 Another name for a mole is: _____

   a nodule.

   b wart.

   c nevus.

   d papule.

   e port-wine stain.

22 Nails are really modified epidermal cells.   _____

    a true

    b false

23 Baldness is partially determined by heredity and _____.   _____

24 Skin color is largely determined by the presence or absence of _____.   _____

25 When sebaceous glands become enlarged due to accumulated sebum,   _____
       _____ often develop.

Bones and Bone Tissue

6

Prefixes

cancel-	lattice; crossbar
chondro-	cartilage
condyl-	knob
intra-	within
lacun-	space; hollow
lamell-	leaf; layer
osteo-	bone
peri-	around
pro-	first
troph-	to nourish

Suffixes

-blast	germ; budding
-clast	to break
-cyte	cell
-genic	to produce
-ostium	door; opening
-physis	to generated
-potent	power

Read Chapter 6 in the textbook, focusing on the major objectives. As you progress through it, with your textbook open, answer and/or complete the following. When you complete this exercise, you will have a thorough outline of the chapter.

DEVELOPING YOUR OUTLINE

I. Introduction (p. 129)

1 Describe osseous (bone) tissue.

2 List several functions of bones.

II. Types of Bones and Their Mechanical Functions (p. 129)

OBJECTIVE 1

1 What are the five classifications of bones, according to shape?

a

b

c

d

e

2 Describe the structure and function of a typical long bone.

3 Describe the structure and function of a typical short bone.

4 Describe the structure and function of a typical flat bone.

5 Describe the structure and function of some typical irregular bones.

6 Describe the structure and function of some typical sesamoid bones.

7 Where are the accessory bones commonly found?

III. Gross Anatomy of a Typical Long Bone (p. 131)

> **OBJECTIVE 2**

1 What are the gross anatomical parts of a typical long bone that can be seen with the unaided eye?

2 What are the microscopic parts of a typical long bone that can be seen only with the aid of a microscope?

IV. Bone (Osseous) Tissue (p. 132)

> **OBJECTIVE 3**

A Compact Bone Tissue (p. 132)

1 Briefly describe the microscopic anatomy of compact bone. Relate its structure to its function.

B Spongy (Cancellous) Bone Tissue (p. 133)

> **OBJECTIVE 4**

1 Briefly describe the microscopic anatomy of spongy (cancellous) bone.

V. Bone Cells (p. 133)

1 Describe the structure and function(s) of each of the following bone cells:

> **OBJECTIVE 5**

a osteogenic cells

> **OBJECTIVE 6**

b osteoblasts

c osteocytes

d osteoclasts

e bone-lining cells

VI. Developmental Anatomy and Growth of Bones (p. 135)

OBJECTIVE 7

A Intramembranous Ossification (p. 136)

    1 Briefly describe the eight steps in intramembranous ossification.

        a

        b

        c

        d

        e

        f

        g

        h

B Endochondral Ossification (p. 138)

    1 Briefly describe the five steps in endochondral ossification.

        a

        b

        c

        d

        e

OBJECTIVE 8

C Longitudinal Bone Growth After Birth (p. 140)

    1 How do bones grow after birth?

D How Bones Grow in Diameter (p. 140)

    1 How do long bones grow in diameter?

VII. Bone Modeling and Remodeling (p. 140)

1 Describe modeling.

2 Describe remodeling.

VIII. The Effects of Aging on Bones (p. 142)

1 What happens when bones age?

IX. When Things Go Wrong (p. 142)

1 Describe each of the following bone disorders:

a osteogenesis imperfecta

b osteomalacia

c rickets

d osteomyelitis

e osteosarcomas

f Paget's disease

These are the terms you should know before proceeding to the post-test.   **MAJOR TERMS**

osseous (bone) tissue *(p. 129)*	yellow bone marrow *(p. 131)*
long bones *(p. 129)*	red marrow *(p. 131)*
short bones *(p. 129)*	endosteum *(p. 131)*
flat bones *(p. 129)*	periosteum *(p. 131)*
irregular bones *(p. 129)*	nutrient foramen *(p. 132)*
sesamoid bones *(p. 129)*	diploë *(p. 132)*
accessory bone *(p. 129)*	osteons *(p. 132)*
diaphysis *(p. 131*	Haversian systems *(p. 132)*
epiphysis *(p. 131)*	lamellae *(p. 132)*
metaphysis *(p. 131)*	central canals *(p. 132)*
epiphyseal (growth) plate *(p. 131)*	perforating canals *(p. 132)*

spongy (cancellous) bone tissue *(p. 133)*

trabeculae *(p. 133)*

osteogenic cells *(p. 133)*

osteoblasts *(p. 133)*

osteocytes *(p. 135)*

osteoclasts *(p. 135)*

bone-lining cells *(p. 135)*

ossification *(p. 136)*

intramembranous ossification *(p. 136)*

center of osteogenesis *(p. 138)*

endochondral ossification *(p. 138)*

primary center of ossification *(p. 138)*

secondary centers of ossification *(p. 140)*

modeling *(p. 140)*

remodeling *(p. 141)*

osteogenesis imperfecta *(p. 142)*

osteomalacia *(p. 142)*

rickets *(p. 142)*

osteomyelitis *(p. 142)*

osteoporosis *(p. 142)*

osteogenic sarcomas *(p. 143)*

Paget's disease *(p. 143)*

RADIATION THERAPY (X-RAY) TECHNOLOGIST AND SURGICAL TECHNOLOGIST

The people who operate radiologic equipment and take x-rays (radiographs) are called **radiation therapy technologists** or **radiographers.** They often work in hospitals and clinics under the supervision of radiologists (physicians who specialize in the use of radiographs). Radiation therapy technologists take x-rays that the radiologist examines to help diagnose a patient's problem. The technologist also administers radiation therapy, and specializes in the use of other radiologic equipment that helps radiologists diagnose and treat illness or injuries.

Radiation therapy technologists must complete a formal educational program in radiography. Programs vary in length from two to four years, and some colleges offer a bachelor's degree in radiologic technology. Some related occupations and specialties include dental hygienists, electrocardiograph technicians, electroencephalographic technologists, and medical technologists.

Surgical technologists work principally in the operating room. They perform functions and tasks that provide a safe environment for surgical care and contribute to the efficiency of the operating room team. The technologists also support the operating surgeons, nurses, and others involved in operative procedures. Knowledge of and experience with aseptic surgical techniques qualify surgical technologists to prepare instruments and materials for use at the operating table and elsewhere and to assist in the use of these materials.

Surgical technologists often work in other patient service settings that call for special knowledge of asepsis. There is great variation in the role of the surgical technologist, depending on the geographic, sociologic, and economic factors of the many different work environments. A high school diploma or equivalent is a prerequisite for all programs in surgical technology.

Based on terminology contained within this chapter, label the following figures: **LABELING ACTIVITY**

1   Osteon.

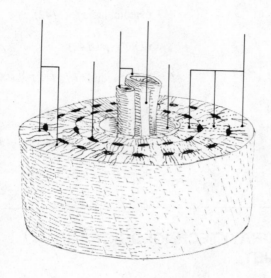

2   Long bone.

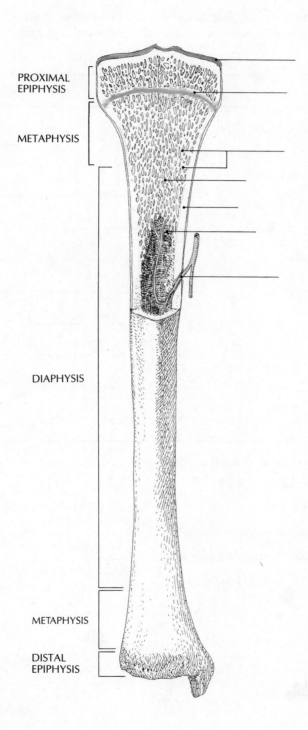

PROXIMAL
EPIPHYSIS

METAPHYSIS

DIAPHYSIS

METAPHYSIS

DISTAL
EPIPHYSIS

After you have completed the following activities, compare your answers with those given at the end of this manual.

1 Match the morphological categories of cartilage and bone tissue with their general characteristics by placing an X in the appropriate space (more than one characteristic may be correct).

	Cartilage	Bone
Derived from embryonic mesenchyme.	_____	_____
Cells housed in lacunae.	_____	_____
Cells form collagen.	_____	_____
Vascular tissue present.	_____	_____
Nutrients reach cells by diffusion.	_____	_____
Tissue can be calcified.	_____	_____
Tissue can regenerate.	_____	_____
Chondroitin sulfate present.	_____	_____
Chondrocytes present.	_____	_____
Fibrils embedded in matrix.	_____	_____

2 Flat (membranous) bones exhibit all of the following *except:* _____

   a they develop by endochondral ossification.

   b they contain bone marrow.

   c they develop by intramembranous ossification.

   d they grow as a result of epiphyseal plate activity.

   e they are composed of both spongy and compact bone.

3 A diploë is: _____

   a an area of spongy bone.

   b found in flat bones.

   c found in irregular bones.

   d a and b.

   e a, b, and c.

4 The osteon (Haversian system) is:

    a the functional unit of compact bone.

    b found in spongy bone.

    c an area that does not change.

    d the area that forms the lamellae.

    e all of the above.

5 The osteon:

    a is composed of concentric lamellae.

    b is found in flat bones.

    c contains periosteal perforating (Sharpey's) fibers.

    d a and **b.**

    e a, **b,** and **c.**

6 Bone lamellae:

    a contain mostly intercellular matrix.

    b contain calcified material.

    c are penetrated by canaliculi.

    d contain collagen fibrils.

    e all of the above.

7 Lamellae are found in the intercellular matrix of adult spongy bone.

    a true

    b false

8 Endochondral ossification is:

    a found in the epiphyses of long bones.

    b occurring at the same time as intramembranous ossification.

    c a process of bone formation involving the replacement of cartilage.

    d a and **b.**

    e a, **b,** and **c.**

9 Matching.

    a ___ rickets   [1] A faulty calcification process.

    b ___ osteomyelitis   [2] A bone tumor.

    c ___ osteomalacia   [3] A bone marrow infection.

    d ___ osteosarcoma   [4] Adult rickets.

10 The closure of the epiphyses of long bones involves the union of bone marrow of the diaphysis with that of the epiphysis.

    a true

    b false

11 Secondary centers of ossification in long bones develop in the _____. _____

12 The primary center of ossification in long bones develops in the _____. _____

13 Long bones' growth in diameter is caused by: _____

   a periosteal activity.

   b bone resorption.

   c intramembranous ossification.

   d **a** and **b**.

   e **a**, **b**, and **c**.

14 Long bones' growth in length is caused by the: _____

   a influence of growth hormone.

   b proliferation of cells in the epiphyseal plate.

   c periosteal activity.

   d **a** and **b**.

   e **a**, **b**, and **c**.

15 All of the following are related to bone tissue *except:* _____

   a elastic fibers.

   b lacuna.

   c canaliculi.

   d collagenous fibers.

   e osteocytes.

16 The space in living bone matrix that is occupied by the osteocyte is called the _____

   _____.

17 The connective tissue that covers the external surface of bone is called the _____

   _____.

18 Matching. _____

   a ___ mucopolysaccharide   [1] Connective tissuecovering of cartilage. _____

   b ___ cartilage growth   [2] Organic ground substance of

                                            cartilage. _____

   c ___ hydroxyapatite   [3] Appositional growth. _____

   d ___ perichondrium   [4] Inorganic bone matrix. _____

19 In the embryo, bone-forming cells are known as _____. _____

20 The deposition of newly mineralized bone is usually complete by the age of 20.

   a true

   b false _____

21 Bones can be classified by their _____ and _____. _____

22 Matching.

   a ___ long bone   [1] Have the same dimensions.   _____

   b ___ short bone   [2] Consist of tables.   _____

   c ___ flat bone   [3] Vertebrae.   _____

   d ___ irregular bone   [4] Length greater than width.   _____

23 List the five types of bone cells.

   a   _____

   b   _____

   c   _____

   d   _____

   e   _____

24 What three minerals are stored in bones?

   a   _____

   b   _____

   c   _____

25 Red bone marrow is also known as _____ tissue.   _____

The Axial Skeleton

7

Prefixes

cervi-	the neck
crib-	sieve
crist-	ridge
ethm-	a sieve
foss-	ditch
fov-	pit
lacr-	tears; weeping
lumb-	the loin
meat-	passage
sphen-	wedge
zyg-	paired

Suffixes

-vertebral	vertebra

Read Chapter 7 in the textbook, focusing on the major objectives. As you progress through it, with your textbook open, answer and/or complete the following. When you complete this exercise, you will have a thorough outline of the chapter.

DEVELOPING YOUR OUTLINE

OBJECTIVE 1

I. Introduction (p. 146)

1 List five functions of the human skeleton.

a

b

c

d

e

II. General Features and Surface Markings of Bones (p. 146)

1 Complete the following table on general surface features of bones (a hint for remembering the markings—all terms that begin with a *T* are projections and those with an *F* are depressions).

Marking	Description	Example
Processes where a bone forms a joint		
condyle	_____	_____
trochlea	_____	_____
facet	_____	_____
head	_____	_____
Processes that are sites of muscle attachment		
crest	_____	_____
spine	_____	_____
trochanter	_____	_____
tubercle	_____	_____
tuberosity	_____	_____
Openings in bones		
fissure	_____	_____
foramen	_____	_____
canal	_____	_____
Depressions commonly seen		
fossa	_____	_____
sulcus	_____	_____
sinuses	_____	_____
notch	_____	_____

III. Divisions of the Skeleton (p. 149)

OBJECTIVE 3

1 What are the two major parts of the skeleton?

a

b

IV. The Skull (p. 150)

OBJECTIVE 4

A Sutures and Fontanels (p. 150)

1 Complete the following table on the major skull sutures:

Suture	Location
coronal	a _____
b _____	between parietal and occipital bones
c _____	between right and left parietal bones
squamous	d _____

2 Complete the following table on the major skull fontanels:

Fontanel	Location
frontal	a _____
b _____	between the occipital and two parietal bones
c _____	between the frontal, parietal, temporal, and sphenoid bones
mastoid	d _____

3 What is a foramen?

B Bones of the Cranium (p. 151)

OBJECTIVE 5

1 How many bones make up the cranium?

2 Complete the following table on the bones of the cranium:

Name	Description
a _____	forms the forehead
b _____	forms most of the superior and lateral wall of the cranium
c _____	contain the styloid process
occipital	**d** _____
sphenoid	**e** _____
f _____	forms the roof of the nasal cavity

3 What is another name for sutural bones?

C Paranasal Sinuses (p. 160)
   1 How are the paranasal sinuses named?

   2 What are two functions of the paranasal sinuses?
    a

    b

D Bones of the Face (p. 161)
   1 Complete the following table on the bones of the face:

Name	Description
a _____	lower jawbone
b _____	form the upper jawbone
c _____	form the posterior part of the hard palate
d _____	cheekbones
e _____	contains a groove that serves as a passageway for tears
nasal	**f** _____
vomer	**g** _____
h _____	project out from the lateral walls of the nasal cavity

E Ossicles of the Ear (p. 163)

    1 Name the three ear ossicles.

        a

        b

        c

F Hyoid Bone (p. 164)

    1 What is unique about this bone?

    2 What is its function?

V. The Vertebral Column (p. 165)

OBJECTIVE 7

    1 What are four functions of the vertebral column?

        a

        b

        c

        d

    2 Complete the following table on the number of adult vertebrae:

Name	Number
a _____	7 _____
thoracic (chest)	**b** _____
c _____	5 _____
d _____	5 fused
e _____	5 fused

A Curvatures of the Vertebral Column (p. 165)

    1 List the four curves of the vertebral column.

        a

        b

        c

        d

    2 What are the functions of these curves?

B A Typical Vertebra (p. 165)

    1 Complete the following table on some common structural features of a typical vertebra:

Feature	Description
a _____	central part of vertebra
b _____	arch
vertebral foramen	**c** _____
spinous process	**d** _____
transverse processes	**e** _____

C Cervical Vertebrae (p. 169)

    1 What is the name of the first two vertebrae?

    2 The cervical vertebrae are referred to as _____ to _____ .

D Thoracic Vertebrae (p. 170)

    1 The thoracic vertebrae are referred to as _____ to _____ .

    2 The ribs articulate with the bodies and _____ processes of the thoracic vertebrae.

E Lumbar Vertebrae (p. 170)

    1 The lumbar vertebrae are referred to as _____ to

    _____.

    2 A lumbar puncture (for examination of spinal fluid) or saddle block

    anesthesia is usually done between lumbar vertebrae _____

    and _____.

F Sacrum and Coccyx (p. 170)

    1 The sacrum is formed by the fusion of _____ vertebrae.

    2 Another name for the tailbone is the _____.

VI. The Thorax (p. 172)

OBJECTIVE 8

    1 List the structures that comprise the thorax.

    a

    b

    c

    d

    2 What are four functions of the thorax?

    a

    b

    c

    d

A Sternum (p. 172)

    1 What are the three portions of the sternum?

    a

    b

    c

B Ribs (p. 172)

OBJECTIVE 9

    1 People have _____ pairs of ribs, all of which articulate

    posteriorly with the _____.

    2 The true ribs attach directly to the _____ and

    _____.

    3 Rib numbers _____ to _____ are known as the

    false ribs.

    4 The floating ribs are numbered _____ and _____.

    5 The space between the ribs is called the _____.

    6 List the major anatomical parts of a typical rib.

      a

      b

      c

      d

      e

VII. When Things Go Wrong (p. 175)

OBJECTIVE 10

  A Fractures (p. 175)

    1 Match the following types of fractures with their descriptions.

a ___ greenstick

b ___ comminuted

c ___ complete

d ___ incomplete

e ___ spiral

f ___ transverse

[1] Bone is broken by twisting.

[2] The distal part of the fibula and the medial
    malleolus are broken.

[3] The distal end of the radius is displaced.

[4] Bone is both broken and bent.

[5] Bone breaks at right angle to axis.

[6] One bone is driven into another.

g ___ oblique   [7] Bone is splintered into small pieces.

h ___ impacted   [8] Skull is broken in a line.

i ___ linear skull   [9] Skull is broken by a puncture.

j ___ depressed skull   [10] The bone does not break completely

                                            into two or more pieces.

k ___ Colles'   [11] Bone is broken at a 45-degree angle.

l ___ Pott's   [12] Bone breaks completely in two places.

B Fractures of the Vertebral Column (p. 177)

    1 Under this category, the most common type of fracture is a

    _____ .

    2 A whiplash injury is likely to cause what type of fracture?

    _____ .

C Herniated Disk (p. 178)

    1 A herniated disk (also called _____ or _____

    disk) occurs when the pulpy center (_____ pulposus)

    protrudes through the outer ring (_____ fibrosus) and pushes

    against a _____ nerve.

D Describe a cleft palate. (p. 179

E Hydrocephalus (p. 179)

    1 Contrast hydrocephalus and microcephalus.

F Spina Bifida (p. 179)

    1 Why is this condition referred to as a cleft spine?

G Spinal Curvatures (p. 179)

    1 Sketch the abnormality indicated by each of the following:

        a Kyphosis   **b** Lordosis   **c** Scoliosis

These are the terms you should know before proceeding to the post-test. **MAJOR TERMS**

processes *(p. 146)*

axial skeleton *(p. 149)*

appendicular skeleton *(p. 149)*

skull *(p. 150)*

foramina *(p. 150*

sutures *(p. 150)*

fontanels *(p. 150)*

calvaria *(p. 151)*

frontal bone *(p. 154)*

parietal bones *(p. 154)*

occipital bone *(p. 154)*

temporal bones *(p. 155)*

sphenoid bone *(p. 160)*

ethmoid bone *(p. 160)*

sutural (Wormian) bones *(p. 160)*

paranasal sinuses *(p. 160)*

maxillary sinuses *(p. 161)*

sphenoid sinuses *(p. 161)*

frontal sinuses *(p. 161)*

inferior nasal conchae *(p. 161)*

vomer *(p. 161)*

palatine bones *(p. 163)*

maxillae *(p. 163)*

zygomatic bones *(p. 163)*

nasal bones *(p. 163)*

lacrimal bones *(p. 163)*

mandible *(p. 163)*

hyoid bone *(p. 164)*

ossicles *(p. 164)*

malleus *(p. 164)*

incus *(p. 164)*

stapes *(p. 164)*

vertebral column *(p. 165)*

vertebrae *(p. 165)*

intervertebral disks *(p. 165)*

atlas *(p. 169)*

axis *(p. 170)*

lumbar puncture *(p. 170)*

sacrum *(p. 170)*

coccyx *(p. 171)*

thorax *(p. 172)*

sternum *(p. 172)*

manubrium *(p. 172)*

xiphoid process *(p. 173)*

ribs *(p. 173)*

true rib *(p. 173)*

false rib *(p. 173)*

floating rib *(p. 173)*

typical rib *(p. 173)*

fracture *(p. 175)*

compression fracture *(p. 177)*

extension fracture *(p. 177)*

herniated disk *(p. 178)*

cleft palate *(p. 179)*

microcephalus *(p. 179)*

hydrocephalus *(p. 179)*

spina bifida *(p. 179)*

kyphosis *(p. 179)*

lordosis *(p. 179)*

scoliosis *(p. 179)*

Based on the terminology contained within this chapter, label the following figures: **LABELING ACTIVITY**

1   Exploded view of the 8 bones in the cranial skull and 21 bones of the facial skull.

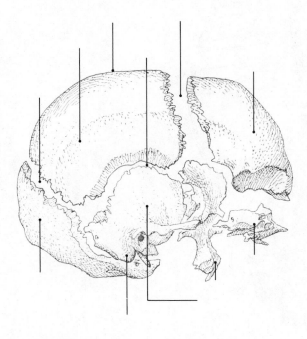

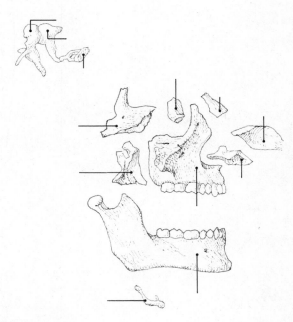

CRANIAL SKULL

FACIAL SKULL

2   Fetal skull.

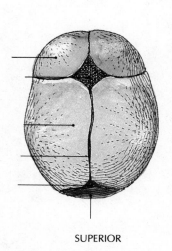

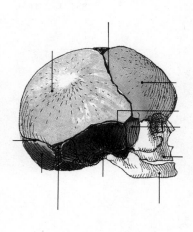

SUPERIOR

LATERAL

3   Vertebral column.

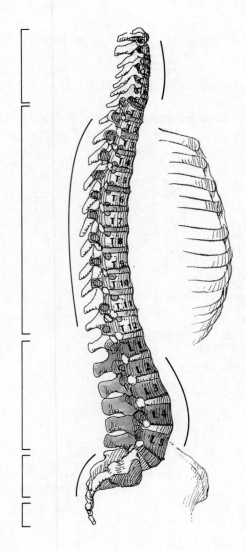

RIGHT LATERAL

4   A typical vertebra.

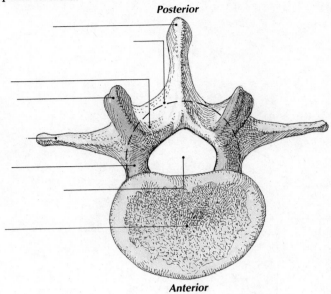

Posterior

Anterior

5   Lumbar vertebrae.

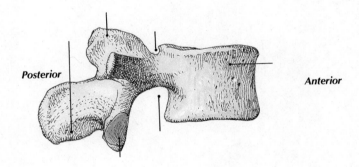

Posterior

Anterior

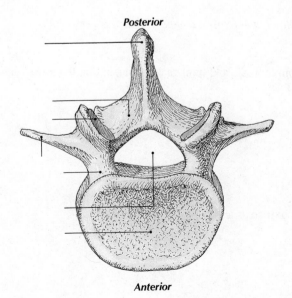

Posterior

Anterior

6   The thorax.

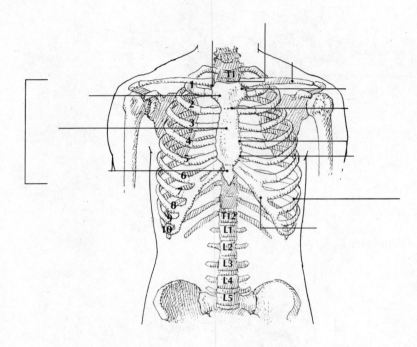

After you have completed the following activities, compare your answers with those given at the end of this manual.   **POST-TEST**

1 Cervical vertebrae differ from thoracic and lumbar vertebrae in that the former have: _____

   a lamina.

   b a pedicle.

   c transverse foramen.

   d transverse processes.

   e both **a** and **b**.

2 The zygomatic process of maxillae articulates with the: _____

   a zygomatic bones.

   b ethmoid bones.

   c sphenoid bones.

   d maxillae and palatine bones.

   e temporal bones.

3   A cleft palate results from the improper fusion of the _____ bone.   _____

    a   mandible

    b   maxilla

    c   ethmoid

    d   navicular

    e   occipital

4   A small, shallow depression is called a: _____

    a   fossa.

    b   notch.

    c   sinus.

    d   meatus.

    e   fovea.

5   A hole or perforation through a bone is called a _____ .   _____

6   True ribs attach via their costal cartilage to the sternum.   _____

    a   true

    b   false

7   The hyoid bone is attached to the skull.   _____

    a   true

    b   false

8   The vertebral column in humans consists of 24 vertebrae.   _____

    a   true

    b   false

9   What are the five divisions of the spinal column?   _____

    a   _____

    b   _____

    c   _____

    d   _____

    e

10   Matching.

    a ___   mandible   [1] Located at the corner of each orbit (eye).   _____

    b ___   frontal bone   [2] Upper jawbone.   _____

    c ___   lacrimal bone   [3] Lower jawbone.   _____

    d ___   vomer   [4] Forms the roof and upper sides of   _____

                                        the skull.

    e ___   ethmoid bone   [5] Cheekbones.   _____

    f ___   zygomatic bone   [6] Forms lower portion of the nasal septum.   _____

g ___ nasal bone	[7] Supports the bridge of the nose.	_____
h ___ temporal bone	[8] Forms the forehead.	_____
i ___ occipital bone	[9] Bone near the ear that connects to lower jaw.	_____
j ___ maxilla	[10] Forms roof of nasal cavity and septum.	_____
k ___ parietal bone	[11] Forms the back and base of the skull.	_____
l ___ sphenoid bne	[12] Butterfly-shaped bone.	_____

11 How many ribs articulate directly with the sternum?   _____

   a 3

   b 5

   c 7   _____

   d 9

   e 10

12 All of the ribs that do *not* articulate directly with the sternum are: _____

   a false ribs.

   b true ribs.

   c floating ribs.

   d sternal ribs.

   e both **a** and **b.**

13 How many vertebrae are located in the thoracic region of the vertebral column?   _____

   a 2

   b 4

   c 8

   d 12

   e 14

14 The most superior part of the sternum is the: _____

   a xiphoid process.

   b manubrium.

   c sternum.

   d sternal angle.

   e body.

15 The neural arch of a vertebra consists of the:

   a pedicle and transverse process.

   b pedicle and lamina.

   c lamina and spine.

   d lamina and spinous process.

   e transverse process and spinous process.

16 Which of the following bones is *not* part of the axial skeleton?

   a ribs

   b vertebrae

   c hyoid bone

   d sternum

   e clavicle

17 Which of the following bones is a component of the axial skeleton?

   a sacrum

   b ilium

   c pubis

   d ischium

   e all of the above

18 The only movable bone of the skull is the:

   a lacrimal.

   b ethmoid.

   c mandible.

   d sphenoid.

   e zygomatic.

The Appendicular Skeleton

8

Prefixes

carp-   wrist
malle-   a hammer
meta-   beyond; change

Suffixes

-osseous   bone; bony

Read Chapter 8 in the textbook, focusing on the major objectives. As you progress through it, with your textbook open, answer and/or complete the following. When you complete this exercise, you will have a thorough outline of the chapter.

DEVELOPING YOUR OUTLINE

I. Introduction (p. 184)

    1 What bones make up the appendicular skeleton?

II. The Upper Extremities (Limbs) (p. 184)

    1 What bones make up the upper extremity?

OBJECTIVE 1

    A Pectoral Girdle (p. 184)

        1 What bones comprise the pectoral girdle?

          a

          b

        2 The pectoral girdle is designed for _____ .

        3 Match the following markings of a clavicle:

    a ___ sternal end   **[1]** Coracoclavicular ligament attaches.

    b ___ acromial end   **[2]** Articulates with the manubrium.

    c ___ costal tuberosity   **[3]** Articulates with the acromion process.

    d ___ coronoid tubercle   **[4]** Costoclavicular ligament attaches.

        4 What is the main function of the clavicle?

        5 On the scapula, what is the function of the acromion?

6 The head of the humerus fits into the _____ of the scapula.

B Bones of the Arm, forearm, and Hand (p. 184)

    1 From the shoulder to the tips of the fingers, list the names of the bones of the upper extremity.

        a

        b

        c

        d

        e

        f

    2 Complete the following table on the bones of the upper extremity:

Bone(s)	Marking	Feature or function
a _____	head	fits into glenoid fossa
humerus	tubercles	**b** _____
humerus	**c** _____	deltoid muscle attaches
humerus	**d** _____	looks like a spool
radius	**e** _____	tendon of the biceps muscle attaches
f _____	coronoid process	grip the trochlea of the humerus
g _____	knuckles	a clenched fist

    3 When the body is in the anatomical position, the _____ is the _____ lateral forearm bone.

4 In the wrist, the proximal carpal bones from lateral to medial position are the:

   a

   b

   c

   d

5 In the wrist, the distal carpal bones from lateral to medial position are the:

   a

   b

   c

   d

6 The five _____ bones make up the hand and the _____ phalanges are the finger bones.

7 The anatomical parts of each phalanx are the:

   a

   b

   c

III. The Lower Extremities (Limbs) (p. 190)

OBJECTIVE 2

   1 From the hips to the tips of the toes, list the names of the major lower extremity bones.

      a

      b

      c

d

e

f

g

A Pelvic Girdle and Pelvis (p. 190)

   1 What bones make up the pelvis?

     a

     b

     c

   2 What are the three parts of each hipbone?

     a

     b

     c

   3 Complete the following table on the bones of the pelvic girdle:

Bone	Marking or Feature	Purpose
ilium, ischium, pubis	a _____	receives the head of the femur
two pubic bones	b _____	forms cartilage joint
c _____	greater sciatic notch	allows blood vessels and sciatic nerve to pass through
pubis	d _____	allows blood vessels and nerves to pass to anterior portion of thigh
ischium	e _____	receives body weight when sitting

4 What are three major functions of the pelvis?

a

b

c

5 Obstetricians often refer to the lesser pelvis in the female as the

_____ pelvis.

6 Complete the following table on the structural differences between the
female and the male pelves:

Characteristics	Female pelvis	Male pelvis
General structure	**a** _____	massive
b _____	widely separated	less widely separated
c _____	face more anteriorly	face laterally
pelvic inlet	circular	**d** _____
pubic arch	obtuse angle	**e** _____
pelvic outlet	**f** _____	narrow

B Bones of the Thigh, Leg, and Foot (p. 193)

1 Complete the following table on the specific features of the femur, tibia,
and fibula:

Feature	Description
Femur	
a _____	The rounded end that fits into the acetabulum.
b _____	Connects the head with the shaft.
c _____	The large lateral process just below the neck.
d _____	The small medial process just below the neck.
linea aspera	**e** _____
condyles	**f** _____
g _____	Prominences on the lateral surfaces of the condyles.

Feature	Description
Tibia	
h _____	A sharp longitudinal ridge on the anterior surface.
i _____	Articulates with the condyles of the femur.
j _____	Articulates with the talus.
Fibula	
k _____	Articulates with the lateral condyle of the tibia.
lateral malleolus	l _____

2 What sesamoid bone of the lower extremity is located within the quadriceps

femoris tendon? _____.

3 Another name for the shinbone is the _____.

4 The _____, or ankle, is composed of seven tarsal bones.

5 Body weight is mostly carried by the two largest tarsals, the

_____, or heelbone, and the _____ that lies

between the tibia and the calcaneus.

6 The bones of the foot include _____ metatarsals, which form

the instep, and _____ phalanges, which form the toes. Each

toe has _____ phalanges except the big toe, which has

_____.

C Arches of the Foot (p. 198)

1 What are two functions of the foot arches?

a

b

2 What are the two longitudinal arches of the foot called?

a

b

3 Weak arches are referred to as _____.

IV. **When Things Go Wrong** (p. 199)

1 Describe shin splints.

These are terms you should know before proceeding to the post-test.   **MAJOR TERMS**

appendicular skeleton *(p. 184)*

upper extremity *(p. 184)*

pectoral (shoulder) girdle *(p. 184)*

clavicle *(p. 184)*

scapula *(p. 184)*

humerus *(p. 186)*

ulna *(p. 187)*

radius *(p. 187)*

carpus (wrist) *(p. 188)*

metacarpus *(p. 188)*

metacarpal bones *(p. 190)*

phalanges *(p. 190)*

pelvic girdle *(p. 190)*

os coxa *(p. 190)*

pelvis *(p. 190)*

greater/false pelvis *(p. 190)*

lesser/true pelvis *(p. 190)*

ilium *(p. 190)*

ischium *(p. 190)*

pubis *(p. 190)*

femur *(p. 193)*

patella *(p. 195)*

tibia *(p. 195)*

fibula *(p. 195)*

tarsus *(p. 195)*

metatarsus *(p. 198)*

phalanges *(p. 198)*

longitudinal arches *(p. 198)*

Bases on the terminology contained within this chapter, label the following figures: **LABELING ACTIVITY**

1   Clavicle and scapula.

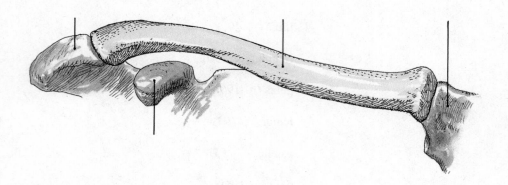

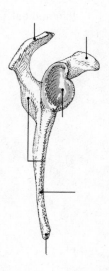

LATERAL

2   Right humerus.

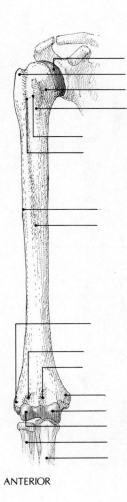

ANTERIOR

3   Right radius and ulna.

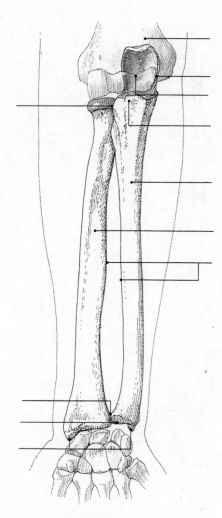

ANTERIOR

4   Bones of the hand.

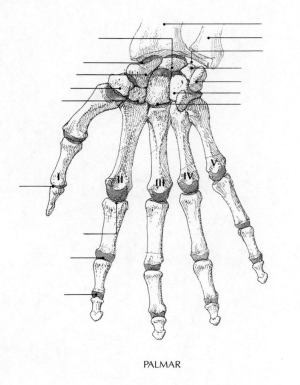

CARPAL BONES

METACARPAL BONES

PHALANGES

PALMAR

5   Right hipbone.

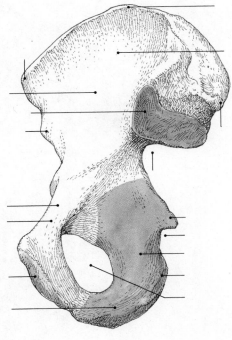

MEDIAL

6   Right femur.

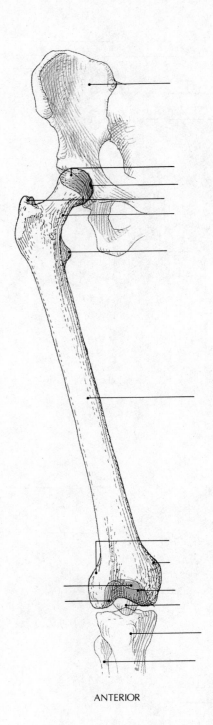

ANTERIOR

7   Right tibia and fibula.

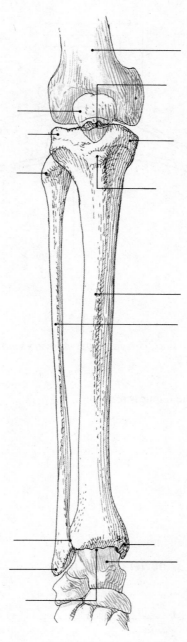

ANTERIOR

8   Bones of the right foot.

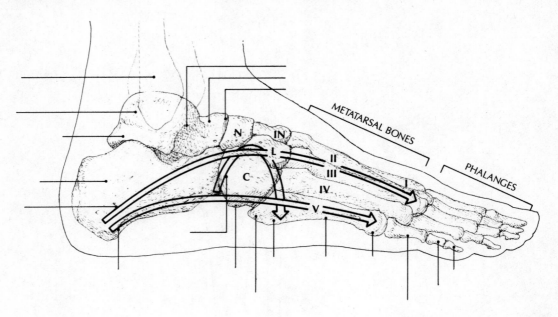

After you have completed the following activities, compare your answers with those given at the end of this manual.

POST-TEST

_____

1 A large, blunt process found only on the femur is a:

  a trochanter.

  b crest.

  c spine.

  d tuberosity.

  e head.

2 Can the following bony landmarks be felt on yourself?

Landmark	Answer (yes or no)	
a sesamoid bone	**a** _____	_____
styloid process of the ulna	**b** _____	_____
tibial tuberosity	**c** _____	_____
head of femur	**d** _____	_____
anterior superior iliac spine	**e** _____	_____
lateral malleolus	**f** _____	_____
mastoid process	**g** _____	_____
radial tuberosity	**h** _____	_____

3 Complete the following table on the structural differences between the male and female pelves:

Feature	Female	Male
bones	smaller	a _____
symphysis pubis	b _____	narrow
obturator foramen	triangular	c _____
iliac fossa	d _____	deep

4 There are five metatarsal bones in the human foot.   _____

  a true

  b false

5 Matching.

  a ___ shoulder bone   [1] metatarsals   _____

  b ___ foot bones   [2] fibula   _____

  c ___ upper arm bones   [3] carpals   _____

  d ___ smaller of the two leg bones   [4] vertebral column   _____

  e ___ breastbone   [5] patella   _____

  f ___ shin bone   [6] tibia   _____

  g ___ thigh bone   [7] scapula   _____

  h ___ kneecap   [8] humerus   _____

  i ___ finger bones   [9] sternum   _____

  j ___ backbone   [10] phalanges   _____

  k ___ hand bones   [11] femur   _____

  l ___ medial lower arm bone   [12] metacarpals   _____

  m ___ lateral lower arm bone   [13] ulna   _____

  n ___ collarbone   [14] radius   _____

  o ___ wrist bones   [15] clavicle   _____

  p ___ hipbone   [16] os coxa   _____

6 The process of the scapula that is shaped like a bent finger is the: _____

  a acromion process.

  b coracoid process.

  c supraglenoid process.

  d spinous process.

  e supraspinatus fossa.

7 Of the following, which is *not* part of the upper extremity?

    a metacarpals

    b humerus

    c tibia

    d carpals

    e radius

Articulations

9

Prefixes

ab-	off; away
ad-	to; toward
art-	joint
bi-	two
burs-	bag; sac
circum-	around
condyl-	knob
duc-	lead
glen-	joint socket
multi-	many
odon-	tooth
ov-	egglike
plant-	sole of foot
phy-	grow; produce
sutur-	sewing
syn-	with; together
uni-	one

Suffixes
none

Read Chapter 9 in the textbook, focusing on the major objectives. As you progress through it, with your textbook open, answer and/or complete the following. When you complete this exercise, you will have a thorough outline of the chapter.

DEVELOPING YOUR OUTLINE

I. Introduction (p. 203)

1 Define the term *articulation*.

OBJECTIVE 1

II. Classification of Joints (p. 203)

1 What are the two methods of classifying joints?

a

b

2 What are three classes of joints based on function? Describe each one.

  a

  b

  c

3 What are the three classes of joints based on structure? Describe each one.

  a

  b

  c

III. Fibrous Joints (p. 203)

OBJECTIVE 1

  1 Describe a fibrous joint.

  2 What are the three types of fibrous joints?

  a

  b

  c

A Sutures (p. 203)

  1 Describe a suture.

  2 Where are sutures found?

B Syndesmoses (p. 203)

  1 Describe a syndesmosis joint.

  2 What determines the amount of movement in this type of joint?

  3 Where would you find a syndesmosis joint?

C Gomphoses (p. 203)

  1 Describe a gomphosis joint.

  2 Where would you find a gomphosis joint?

IV. Cartilaginous Joints (p. 203)

OBJECTIVE 3

  1 Describe a cartilaginous joint.

A Synchondroses (p. 206)

  1 Describe a synchondrosis joint.

  2 What is the main function of a synchondrosis joint?

B Symphyses (p. 206)

  1 Describe a symphysis joint.

  2 Where would you find a symphysis joint?

V. Synovial Joints (p. 206)

OBJECTIVE 4

A Typical Structure of Synovial Joints (p. 206)

  1 What are four essential features of a synovial joint?

    a

    b

    c

    d

B Bursae and Tendon Sheaths (p. 208)

OBJECTIVE 5

  1 What is the structure of bursae?

  2 What are the functions of tendon sheaths?

VI. Movement at Synovial Joints (p. 208)

1 What three factors limit movement at synovial joints?

a

b

c

A Terms of Movement (p. 209)

1 Complete the following table on types of movements at synovial joints:

Action	Definition
a _____	Angle between two bones is decreased.
b _____	Angle between two bones is increased.
c _____	Movement away from midline of body.
d _____	Movement toward midline of body.
e _____	Thumb touches an opposite finger.
f _____	Forward-pushing movement.
g _____	Lowering a body part.

B Uniaxial, Biaxial, and Multiaxial Joints (p. 209)

1 Where in the body would you find each of the following types of joints?

a uniaxial _____

b biaxial _____

c multiaxial _____

2 Which of the above joints exhibits the greatest degree of freedom with

respect to movement?

VII. Types of Synovial Joints (p. 209)

1 Complete the following table on the synovial joints:

Type	Axes of motion permitted	Example
gliding	biaxial	a _____
b _____	triaxial	hip, shoulder
c _____	biaxial	wrist
saddle	d _____	thumb joint
e _____	uniaxial	atlantoaxial joint
hinge	f _____	elbow

VIII. Description of Some Major Joints (p. 213)

OBJECTIVE 10

For each of the following joints, complete the asked-for information.

A Temporomandibular (Jaw) Joint (p. 213)

1 articulation _____

2 classification _____

3 movement _____

B Shoulder Joint (p. 216)

1 articulation _____

2 classification _____

3 movement _____

C Hip Joint (p. 216)

1 articulation _____

2 classification _____

3 movement _____

D Knee Joint (p. 218)

1 articulation _____

2 classification _____

3 movement _____

E Which is more stable, the shoulder or hip joint? Explain.

IX. Nerve Supply and Nutrition of Synovial Joints (p. 220)

A Nerve Supply of Joints (p. 220)

1 What is a sensory nerve fiber?

2 What is a motor nerve fiber?

B Blood and Lymph Supply of Joints (p. 221)

1 Why is physical activity essential for the maintenance of healthy joints?

2 What parts of a synovial joint do not receive nutrients from the blood vessels?

X. Effects of Aging on Joints (p. 221)

1 What happens to joints with age?

XI. When Things Go Wrong (p. 221)

1 Explain why the knee joint is injured so often in active sports.

2 Complete the following table on joint diseases:

Disease	Characteristic
a _____	Means inflammation of a joint.
b _____	Accumulation of uric acid in joints.
c _____	Most common form of arthritis.
d _____	Most debilitating form of chronic arthritis.
e _____	Fibrous tissue develops on the articulating surface of a joint.
bursitis	f _____
g _____	A partial joint dislocation.
bunion	h _____

These are terms you should know before proceeding to the post-test. **MAJOR TERMS**

articulation *(p. 203)*

synarthrosis *(p. 203)*

amphiarthrosis *(p. 203)*

diarthrosis *(p. 203)*

fibrous joints *(p. 203)*

suture *(p. 203)*

syndesmosis *(p. 203)*

gomphosis *(p. 203)*

cartilaginous joints *(p. 203)*

synchondrosis *(p. 206)*

symphysis *(p. 206)*

synovial joints *(p. 206)*

synovial cavity *(p. 206)*

articular (hyaline) cartilage *(p. 206)*

articular disks *(p. 206)*

menisci *(p. 206)*

articular capsule *(p. 206)*

ligaments *(p. 207)*

bursae *(p. 208)*

tendon (synovial) sheath *(p. 208)*

uniaxial joint *(p. 209)*

biaxial joint *(p. 209)*

multiaxial (triaxial) joint *(p. 209)*

hinge joints *(p. 209)*

pivot joints *(p. 209)*

condyloid joints *(p. 209)*

gliding joints *(p. 213)*

saddle joints *(p. 213)*

ball-and-socket joints *(p. 213)*

temporomandibular (TM) joint *(p. 213)*

shoulder joint *(p. 216)*

rotator cuff *(p. 216)*

hip (coxal) joint *(p. 216)*

knee (tibiofemoral) joint *(p. 218)*

PHYSICAL THERAPIST AND PHYSICAL THERAPIST ASSISTANT

Physical therapists evaluate and treat functional disabilities of muscle, joints, or bones that have resulted from disease or injury. They are trained to deliver therapies that help restore strength and mobility to people with these diseases and injuries. Physical therapy both restores physical function and prevents further disability from injury or illness. Physical therapists use exercise, massage, light, water, electricity, and application of heat and cold to relieve pain and improve the condition of muscles, bones, and joints. Sports medicine and sports injury rehabilitation programs incorporate physical therapy.

Some therapists work in hospitals. Others work in nursing homes, rehabilitation centers, schools for handicapped children, and clinics. Some work for public health departments or home health agencies, and others teach or serve as consultants. Many physical therapists work part time.

Physical therapists must be licensed by the state that they practice in. In order to become licensed, each candidate must first earn a bachelor's degree in physical therapy or become certified through a specific physical therapy course (12 to 16 months), then must pass a state board exam. Teaching, research, and administrative positions often require a graduate degree.

Physical therapist assistants work under the supervision of professional therapists to help rehabilitate disabled patients. The assistant can administer treatments that help to restore physical functions and prevent disability from injury or illness. Assistants work with physical therapists to test patients and determine the extent of their capabilities and the best treatment. Using special therapy equipment, they apply heat, cold, light, ultrasound, and massage, and report to their supervisors on how well the patient is responding to treatment. Assistants help patients perform therapeutic exercises such as walking and climbing stairs. They also help physical therapists instruct patients on the use of braces, splints, and artificial limbs.

Training requirements vary, with some states requiring that assistants be licensed. Many colleges offer accredited two-year programs.

Based on the terminology contained within this chapter, label the following synovial joint:

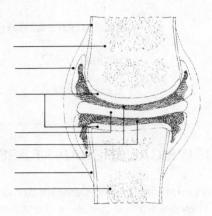

After you have completed the following activities, compare your answers with those given at the end of this manual.

1 Synovial fluid:

   a is a lubricant.

   b is secreted by synovial membranes.

   c provides nutrients for articular cartilage.

   d a and **b**.

   e a, **b**, and **c**.

2 Fibrocartilage is typically found in the:

   a intervertebral disks.

   b ligamentum teres femoris.

   c sternoclavicular joint.

   d symphysis pubis.

   e all of the above.

3 Articular cartilage:

   a is covered by perichondrium.

   b receives its nutrients by direct diffusion from blood.

   c is also elastic cartilage.

   d has its cells arranged in columns.

   e is none of the above.

4 Place a VC by the ligaments associated with the vertebral column, HJ by the ligaments associated with the hip joint, KJ by the ligaments associated with the knee joint, and N for a structure that is not a ligament.

a _____ posterior cruciate   _____

b _____ articular capsule   _____

c _____ medial (tibial) collateral   _____

d _____ iliofemoral   _____

e _____ posterior longitudinal   _____

f _____ arcuate popliteal   _____

5 Matching.

a ___ flexion   [1] Decrease the angle.   _____

b ___ extension   [2] Increase the angle.   _____

c ___ hyperextension   [3] Toward the midline.   _____

d ___ abduction   [4] Away from the midline.   _____

e ___ adduction   [5] Continuation of extension.   _____

6 A joint is defined as a meeting place of three or more bones.   _____

a true
b false

7 Classification of joints depends on the type of connective tissue between the articulating bones.   _____

a true
b false

8 Cartilaginous joints are symphyses.   _____

a true
b false

9 Synchondroses typically allow no movement.   _____

a true
b false

10 A dislocation is the actual displacement of the articulating surfaces of bones.   _____

a true
b false

11 The type of joint found between vertebrae is called a: _____

a synchondrosis.
b symphysis.
c suture.
d syndesmosis.
e none of the above.

12 An example of a hinge joint would be the _____.   _____

13 The class of joints that contains a cavity is the _____.   _____

14 The head of the humerus articulates with the: _____

   a olecranon fossa.

   b glenoid fossa.

   c coronoid fossa.

   d acromion.

   e none of these.

15 Matching.

a ___ gomphosis	[1] Between cranial bones.	_____
b ___ syndesmosis	[2] Between distal ends of the tibia and fibula.	_____
c ___ synchondrosis	[3] A tooth in the mandible.	_____
d ___ ball-and-socket	[4] Symphysis pubis.	_____
e ___ found in upper arm	[5] Epiphyseal plate of a growing bone.	_____
f ___ trochoid	[6] Between the two carpal bones.	_____
g ___ ellipsoidal	[7] Shoulder and hip joints.	_____
h ___ sutures	[8] Radiocarpal joint.	_____
i ___ plane (gliding)	[9] The elbow joint.	_____
j ___ symphysis	[10] Joint between C_1–C_2 vertebrae.	_____

16 Which of the following is *not* involved with reducing friction during movement?   _____

   a bursa

   b collateral ligaments

   c tendon sheath

   d synovial cavity

   e synovial fluid

17 The primary function of bursae and tendon sheaths is to decrease friction.   _____

   a true

   b false

18 A suture is one example of a _____ joint.   _____

19 All joints that contain a synovial cavity are classified as: _____

   a synarthroses.

   b symphyses.

   c synchondroses.

   d sutures.

   e diarthroses.

20 Which of the following types of joints has strong collateral ligaments that restrict   _____

    movement to just one plane?

    a gliding

    b ball-and-socket

    c pivot

    d hinge

    e plane

21 Which of the following joints is most stable?   _____

    a hip

    b TM

    c shoulder

    d knee

    e a toe joint

22 A fluid-filled sac enclosed by a synovial membrane is a: _____

    a coelum.

    b bursa.

    c diarthrosis.

    d vesicle.

    e none of the above.

23 A gliding joint is: _____

    a biaxial.

    b nonaxial.

    c triaxial.

    d multiaxial.

    e none of the above.

24 The primary symptom of rheumatoid arthritis is inflammation of the _____ .   _____

25 The displacement of a bone from its joint is called _____.   _____

Muscle Tissue

10

Prefixes

apo-	from
auto-	within; self
endo-	within
epi-	on
fasc-	bundle
inter-	between
myo-	muscle
peri-	around
rigor-	stiffness
sarco-	flesh; muscle
sub-	below
super-	above
tetan-	rigid; stretched
tri-	three; triple

Suffixes

-dermal	skin
-fibril	small fiber
-lemma	husk; covering
-metric	measurement
-plasm	formed material
-tome	cutting

Read Chapter 10 in the textbook, focusing on the major objectives. As you progress through it, with your textbook open, answer and/or complete the following. When you complete this exercise, you will have a thorough outline of the chapter.

DEVELOPING YOUR OUTLINE

I. Introduction (p. 229)

1 What are the four physiological properties of muscle tissue?

OBJECTIVE 1

a

b

c

d

2 List the three types of muscle tissue and give an example of a function performed by each.

    a

    b

    c

II. Skeletal Muscle (p. 229)

    1 What are two other descriptive names for skeletal muscle?

        a

        b

    2 What is muscle tone?

A Cell Structure and Organization (p. 229)

    1 Complete the following table on muscle cell structure:

Cell part	Function
a _____	Specialized muscle cells.
b _____	Muscle cell membrane.
sarcoplasm	**c** _____
d _____	Individual muscle fibers.
e _____	Network of cytoplasmic tubules.
f _____	Tubes that run at right angles to the sarcoplasmic reticulum.
myosin	**g** _____
h _____	Thin myofilaments.
i _____	The fundamental unit of muscle contraction.
j _____	Area where the sarcoplasmic reticulum and T tubules meet.

    2 What produces the dark and light bands in skeletal muscle?

    3 What is the fundamental unit of muscle contraction?

B Fasciae (p. 231)

    1 What are the two forms of faciae? Describe each one briefly.

       a

       b

OBJECTIVE 4

C Other Connective Tissue Associated with Muscles (p. 231)

    1 Indicate the structure of each of the following tissues that are associated with muscles:

       a epimysium

       b perimysium

       c fascicles

       d endomysium

       e tendon

       f aponeurosis

D Blood and Nerve Supply (p. 232)

    1 Why do muscles need an adequate blood supply?

    2 How can a nerve stimulate several muscle fibers at once?

OBJECTIVE 5

E Nervous Control of Muscle Contractions (p. 232)

    1 Complete the following table on the nervous control of muscle contraction:

OBJECTIVE 6

Structure	Description
a _____	A motor neuron and the muscle fiber it innervates.
b _____	Site where a motor nerve ending contacts a muscle fiber.
axon terminals	**c** _____
motor end plate	**d** _____
e _____	Tiny gap at motor end plate.

III. Smooth Muscle (p. 234)

OBJECTIVE 7

1 Where are four body sites where smooth muscle is commonly found?

a

b

c

d

A Properties of Smooth Muscle (p. 235)

1 What are the two main anatomical characteristics of smooth muscle?

a

b

B Structure of Smooth Muscle Fibers (p. 235)

1 What are eight anatomical differences between smooth and skeletal muscles?

a

b

c

d

e

f

g

h

2 What are the two types of smooth muscle fibers?

a

b

3 What are the two types of contractions that take place in a single-unit smooth muscle?

a

b

4 What is unique about multiunit smooth muscle?

IV. Cardiac Muscle (p. 236)

OBJECTIVE 8

A Structure of Cardiac Muscle (p. 236)

   1 What are the functions of intercalated disks?

a

b

c

d

V. Developmental Anatomy of Muscles (p. 237)

1 How does skeletal muscle undergo development?

2 How does smooth muscle undergo development?

3 How does cardiac muscle undergo development?

VI. The Effects of Aging on Muscles (p. 238)

1 How do muscles age?

VII. When Things Go Wrong (p. 239)

A Muscular Dystrophy (p. 239)

   1 Describe this muscular disorder.

B Myasthenia Gravis (p. 239)

    1 What are some of the defects in myasthenia gravis?

C Tetanus (Lockjaw) (p. 239)

    1 Why is tetanus called lockjaw?

    2 What causes tetanus?

These are terms you should know before proceeding to the post-test.   **MAJOR TERMS**

contractility *(p. 229)*

excitability *(p. 229)*

extensibility *(p. 229)*

elasticity *(p. 229)*

skeletal muscle tissue *(p. 229)*

muscle fibers *(p. 229)*

sarcolemma *(p. 229)*

sarcoplasm *(p. 229)*

myofibrils *(p. 229)*

sarcoplasmic reticulum *(p. 229)*

transverse (T) tubules *(p. 229)*

triad *(p. 229)*

myofilaments *(p. 229)*

myosin *(p. 229)*

actin *(p. 229)*

sarcomere *(p. 231)*

fascia *(p. 231)*

superficial fascia *(p. 231)*

deep fascia *(p. 231)*

epimysium *(p. 231)*

perimysium *(p. 231)*

fascicles *(p. 231)*

endomysium *(p. 231)*

tendon *(p. 231)*

aponeurosis *(p. 231)*

motor unit *(p. 232)*

myoneural (neuromuscular) junction *(p. 232)*

motor end plate *(p. 232)*

synaptic gutter (trough) *(p. 232)*

subneural clefts *(p. 232)*

synaptic cleft *(p. 234)*

single-unit smooth muscle *(p. 235)*

multiunit smooth muscle *(p. 236)*

cardiac muscle tissue *(p. 236)*

intercalated disks *(p. 237)*

mesodermal cells *(p. 237)*

somites *(p. 237)*

dermatome *(p. 237)*

sclerotome *(p. 237)*

myotome *(p. 237)*

OCCUPATIONAL THERAPIST AND OCCUPATIONAL THERAPY ASSISTANT

Occupational therapists organize educational, vocational, and recreational activities that teach mentally or physically disabled people how to be self-sufficient. They evaluate the abilities and skills of patients, set goals, and plan a therapy program together with the patient and members of a medical team that may include physicians, physical therapists, vocational counselors, nurses, social workers, and other specialists.

Occupational therapy helps physically disabled patients to determine the extent of their abilities and their capacity to regain physical, mental, or emotional stability. Occupational therapists teach manual and creative skills, such as weaving and leatherworking, and business and industrial skills, such as typing and the use of power tools. These skills help to restore the patient's mobility, coordination, and confidence. Therapists also plan and direct games and other activities, especially for children. They may design and make special equipment or devices to help disabled patients.

A bachelor's degree and certification in occupational therapy are required. Coursework includes physical, biological, and behavioral sciences and the application of occupational therapy theory and skills. These programs also require students to work for six to nine months in hospitals or health agencies to gain experience in clinical practice.

Personal qualifications needed in the profession include maturity, patience, imagination, manual skills, the ability to teach, and a sympathetic but objective approach to illness and disability.

Occupational therapy assistants work under the supervision of professional occupational therapists to help rehabilitate patients who are physically or mentally disabled. They help plan and implement programs of educational, vocational, and recreational activities that strengthen patients' muscle power, increase motion and coordination, and develop self-sufficiency in overcoming disabilities. Many occupational therapy assistants work in hospitals. Others work in nursing homes, schools for handicapped children and the mentally retarded, rehabilitation and day care centers, special workshops, and outpatient clinics.

Junior and community colleges offer educational programs that prepare occupational therapy assistants. These programs award an associate degree upon completion of vocational or technical courses. Approved programs combine classroom instruction with at least two months of supervised practical experience. Courses include human development, anatomy and physiology, and the effects of illness and injury on patients. Students also practice skills and crafts they will later teach to patients.

Based on the terminology contained within the chapter, label the following muscle cell: **LABELING ACTIVITY**

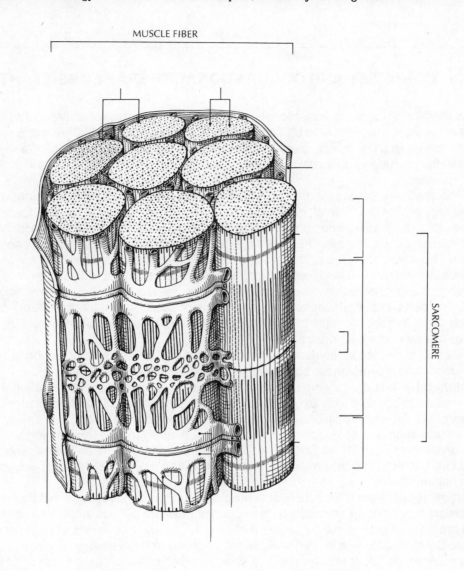

MUSCLE FIBER

SARCOMERE

After you have completed the following activities, compare your answers with those given at the end of this manual.   **POST-TEST**

1 Each skeletal muscle fiber is surrounded by a layer of connective tissue called the: _____

   a epimysium.

   b perimysium.

   c promysium.

   d endomysium.

   e none of the above.

2 In a muscle cell, the proteins found within the thin filaments are: _____

   a actin.

   b troponin.

   c topomyosin.

   d **a** and **b.**

   e **a, b,** and **c.**

3 Which muscle disorder results from the failure of the excitation stimulus to be _____
transmitted across the myoneural junction?

   a myasthenia gravis

   b muscular dystrophy

   c tetanus

   d trichinosis

   e none of the above

4 A contraction in which the cell is not allowed to shorten in length is called: _____

   a isometric.

   b hypermetric.

   c tonic.

   d hypotonic.

   e isotonic.

5 The cross bridges of the myosin filament: _____

   a shorten during the contraction process.

   b are made up of toponin molecules.

   c are attached to ATP molecules that drive the power stroke of the head of the
cross bridge.

   d **a** and **b.**

   e **a, b,** and **c.**

6 Matching.

   a ___ actin   [1] Myofilaments make up a. _____

   b ___ troponin   [2] Contain the protein myosin. _____

   c ___ myofibril   [3] Attached to the Z line. _____

   d ___ smooth muscle   [4] A regulator protein. _____

   e ___ recruitment of motor   [5] A single muscle cell. _____
          neurons

   f ___ cardiac muscle   [6] The combination of the motor _____
                                neuron and muscle fiber.

   g ___ aetylcholine   [7] A response to a stimulation of high _____
                                intensity.

h ___ muscle fiber   [8] Lacks a well-developed sarcoplasm.   _____

i ___ A band   [9] Has a well-developed sarcoplasm.   _____

j ___ motor unit   [10] A neurotransmitter.   _____

7   A skeletal muscle fiber is normally part of only one motor unit.   _____

   a true
   b false

8   The greater the concentration of ACh, the stronger the force of contraction.   _____

   a true
   b false

9   Intestinal smooth muscle contracts spontaneously in the absence of nerve stimulation.   _____

   a true
   b false

10   A single motor neuron and all of the muscle cells it innervates are called a   _____

   _____ .

11   When a skeletal muscle cell contracts, the _____ band and   _____

   _____ zone shorten in length.   _____

12   During a muscle contraction, the   _____

   a I band shortens.

   b sarcomere shortens.

   c distance between two Z lines decreases.

   d a and **b.**

   e a, b, and **c.**

13   The actin protein in a myofibril contains: _____

   a a helical strand of actin.

   b two topomyosin strands.

   c troponin attached at regular intervals along the tropomyosin strand.

   d a and **b.**

   e a, b, and **c.**

14   Which of the following is *not* a basic property of muscle tissue?   _____

   a irritability

   b extensibility

   c excitability

   d contractility

   e elasticity

15   Most smooth muscle is the _____ unit type.   _____

16   One of the most useful by-products of muscle contraction is _____ .   _____

17   The fibrous connective tissue that covers the skeletal muscles and holds them   _____
     together is called _____ .

The Muscular System

11

Prefixes

apo-	from
bi-	two
fusi-	spindle-like
infra-	below
lun-	the moon
multi-	many
peri-	around
perone-	fibula
semi-	half
sphin-	squeeze
syn-	together
uni-	one

Suffixes

-form	shape
-penna	feather
-ineum	discharge

Read Chapter 11 in the textbook, focusing on the major objectives. As you progress through it, with your textbook open, answer and/or complete the following. When you complete this exercise, you will have a thorough outline of the chapter.

DEVELOPING YOUR OUTLINE

I. Introduction (p. 244)

    1 About how many skeletal muscles make up your body?

II. How Muscles Are Named (p. 244)

OBJECTIVE 1

    1 Describe seven ways muscles are named.

        a

        b

        c

        d

        e

f

g

2 For each of the listed muscles, tell what criterion is exemplified by its name.

Muscle name	Criterion
example: intercostal muscle	location
brachii	a _____
tibialis posterior	b _____
rectus abdominus	c _____
abductor polis	d _____
deltoid	e _____
longus	f _____
biceps	g _____
sternocleidomastoid	h _____

III. Attachment of Muscles (p. 244)

OBJECTIVE 2

1 How are muscles attached to bones?

IV. Architecture of Muscles (p. 245)

OBJECTIVE 3

1 Complete the following table of each of the basic patterns of muscle fiber arrangement within fascicles (bundles).

Classification	Description	Example
a _____	All fascicles run parallel to the long axis of the muscle.	b _____
unipennate	c _____	d _____
multipennate	Many oblique fascicles.	e _____
fusiform	f _____	biceps brachii
g _____	Circular fascicles.	h _____

2 Describe the action of the following types of muscles:

   a flexor

   b extensor

   c abductor

   d adductor

   e pronator

   f supinator

   g rotator

   h levator

   i depressor

   j protractor

   k retractor

   l tensor

   m sphincter

   n evertor

   o invertor

V. Individual and Group Actions of Muscles (p. 247)

1 List and describe the four basic roles muscles play in producing or restricting movement.

   a

   b

c

d

2 Complete the table below of muscle types based on action:

Muscle type	Action
a _____	Bending so that the angle between two bones decreases.
extensor	**b** _____
c _____	Turning forearm so that palm faces downward.
levator	**d** _____
e _____	Movement in a forward direction.
f _____	Reduces the size of an opening.

VI. Lever Systems and Muscle Actions (p. 250)

1 What two forces act within every lever system?

a

b

A First-Class Levers (p. 250)

1 Describe this lever system.

2 Give an example of where in the body this system is found.

B Second-Class Levers (p. 250)

1 Describe this lever system.

2 Give an example of where in the body this system is found.

C Third-Class Levers (p. 250)

1 Describe this lever system.

2 Give an example of where in the body this system is found.

D Leverage: Mechanical Advantage (p. 251)

   1 What is leverage?

   2 What is the advantage of leverage?

VII. Specific Actions of Principal Muscles (p. 251)

OBJECTIVE 7

  A Muscles of Facial Expression (p. 252)

    1 Give the name of the facial muscle(s) that accomplish the following:

      a elevates eyebrows

      b wiggles ears

      c closes eyelid

      d closes mouth

      e the "grinning" muscle

    2 Besides being located in the face, the muscles of facial expression are

      also found in the _____ and _____.

    3 What are two unique characteristics of all facial muscles?

      a

      b

    4 What is the correct name for the trumpeter's muscle?

    5 Why is the cliche "it takes more muscles to frown than to smile" true?

  B Muscles That Move the Eyeball (p. 254)

    1 Why are the eyeball muscles considered extrinsic?

    2 With respect to the eyeball muscles, the _____ recti muscles

      insert in front of the horizontal axis and the _____ oblique

      muscles insert behind the horizontal axis.

    3 A disorder in which both eyes cannot be directed at the same point or

      object at the same time is called _____ .

4 Name the muscles of the eyeballs that perform the actions listed below.

   a rolls eye upward

   b rolls eye downward

   c rolls eye laterally

   d rolls eye medially

   e elevates, adducts, and rotates eye

C Muscles of Mastication (p. 255)

   1 List the four pairs of muscles that produce biting and chewing movements.

      a

      b

      c

      d

   2 What muscle of mastication produces side-to-side movements during chewing?

   3 What nerve innervates the muscles of mastication?

   4 Which mastication muscle has its origin on the zygomatic arch?

D Muscles That Move the Hyoid Bone (p. 256)

   1 What four muscles are associated with the floor of the mouth?

      a

      b

c

d

2 What four muscles are known as the strap muscles?

a

b

c

d

E Muscles That Move the Tongue (p. 257)

1 What three muscles enable the tongue to protrude, retract, and depress?

a

b

c

2 What nerve innervates the tongue muscles?

3 What muscle can cause the tongue to block the respiratory passage?

4 What group of muscles is responsible for altering the shape of the tongue?

F Muscles That Move the Head (p. 258)

1 The sternocleidomastoid muscle has its insertion on the

_____ of the temporal bone.

2 The muscle that flexes the head is the _____.

G Intrinsic Muscles That Move the Vertebral Column (p. 259)

    1 What are the two major groups of muscles that make up the intrinsic back muscles?

        a

        b

    2 What are the three columns of muscle bundles of the superficial group of vertebral column muscles?

        a

        b

        c

    3 The spinalis muscles are called _____ or _____, depending on their location.

H Muscles Used in Quiet Breathing (p. 260)

    1 What muscle increases the vertical length and volume of the thorax?

    2 Which muscles elevate the ribs?

    3 Which muscles are attached to the first rib?

I Muscles That Support the Abdominal Wall (p. 262)

    1 What is the name of the tendon that runs from the xiphoid process to the pubic symphysis?

    2 List the four groups of abdominal muscles from the outside to the inside.

        a

        b

        c

        d

3 Which muscle extends from the twelfth rib to the posterior iliac crest?

J Muscles That Form the Pelvic Outlet (p. 264)
    1 The _____ is the funnel-shaped muscular floor of the pelvic cavity.

    2 What specific muscles comprise the pelvic diaphragm?

    3 What diaphragm lies below the pelvic diaphragm?

K Muscles That Move the Shoulder Girdle (p. 266)
    1 Give the action of the following muscles:
        a trapezius

        b levator scapulae

        c pectoralis minor

        d serratus anterior

        e pectoralis major

        f subclavius

L Muscles That Move the Humerus at the Glenohumeral (Shoulder) Joint (p. 268)
    1 List the four SITS muscles.
        a

        b

        c

        d

2 What four muscles flex the humerus?

   a

   b

   c

   d

3 What five muscles extend the humerus?

   a

   b

   c

   d

   e

M Muscles That Move the Forearm and Wrist (p. 270)

   1 Which three muscles flex the forearm at the elbow joint?

   a

   b

   c

   2 What two muscles extend the forearm at the elbow joint?

   a

   b

3 What two muscles pronate the forearm at the radioulnar joint?

   a

   b

4 What two muscles supinate the forearm at the radioulnar joint?

   a

   b

N Muscles That Move the Wrist and Hand at the Radiocarpal and Midcarpal Joints (p. 272)

   1 List the seven muscles that act on the radiocarpal and midcarpal joints.

      a

      b

      c

      d

      e

      f

      g

O Muscles That Move the Thumb (p. 273)

   1 Give the action of the following muscles:

      a flexor pollicis longus

      b opponens pollicis

      c adductor pollicis

      d abductor pollicis longus

P Muscles That Move the Fingers (Except the Thumb) (p. 274)

    1 Give the action of the following muscles:

        a flexor digitorum profundus

        b lumbricals

        c extensor digitorum

        d abductor digiti minimi

Q Muscles That Move the Femur at the Hip Joint (p. 276)

    1 What are the three hamstring muscles?

        a

        b

        c

    2 Give the action of the following muscles:

        a pectineus

        b iliopsoas

        c adductor magnus

        d gluteus maximus

        e gluteus minimus

R Muscles That Act at the Knee Joint (p. 278)

    1 Give the action of the following muscles:

        a rectus femoris

        b vastus lateralis

   c semimembranosus

   d popliteus

S Muscles That Move the Foot at the Talocrural and Subtalar Joints (p. 280)
   1 What two muscles are the plantar flexors?

     a

     b

T Muscles That Move the Foot at the Subtalar Axis (Axis of Henke) (p. 280)
   1 What two muscles invert the axis of Henke?

     a

     b

   2 What three muscles evert the axis of Henke?

     a

     b

     c

U Muscles of the Toes (p. 282)
   1 Complete the following list of the four basic muscle layers of the foot:
   First layer of muscles

     a

     b

     c

   Second layer of muscles

     d

     e

Third layer of muscles

f

g

h

Fourth layer of muscles

i

j

VIII. When Things Go Wrong (p. 287)
    A Hernias (p. 287)
        1 What is the most common type of hernia?

        2 What type of hernia is more common in women than in men?

    B Tendinitis (p. 287)
        1 What three body sites are primarily involved in tendinitis?
          a

          b

          c

    C Tennis Elbow (p. 287)
        1 What is the medical name for tennis elbow?

    D Tension Headache (p. 287)
        1 What part of the head is involved in this type of headache?

These are terms you should know before proceeding to the post-test.

MAJOR TERMS

muscular system *(p. 244)*

belly *(p. 244)*

tendon *(p. 244)*

aponeurosis *(p. 244)*

origin *(p. 244)*

insertion *(p. 244)*

agonist *(p. 250)*

antagonist *(p. 250)*

Based on the terminology contained within this chapter, label the following figures: **LABELING ACTIVITY**

1   Anterior view of the muscular system.

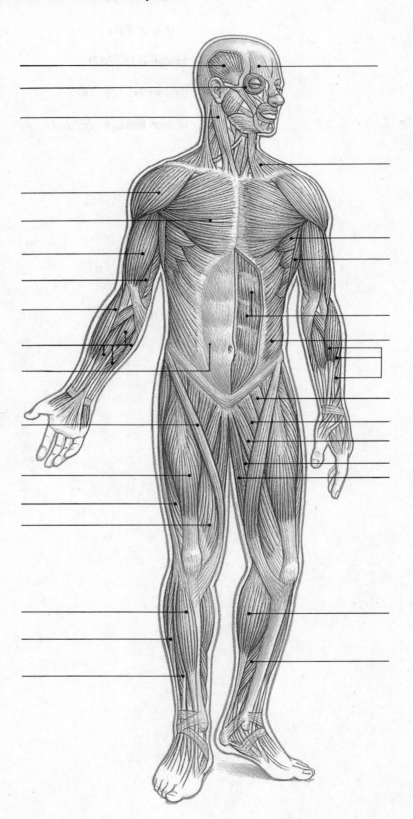

2   Posterior view of the muscular system.

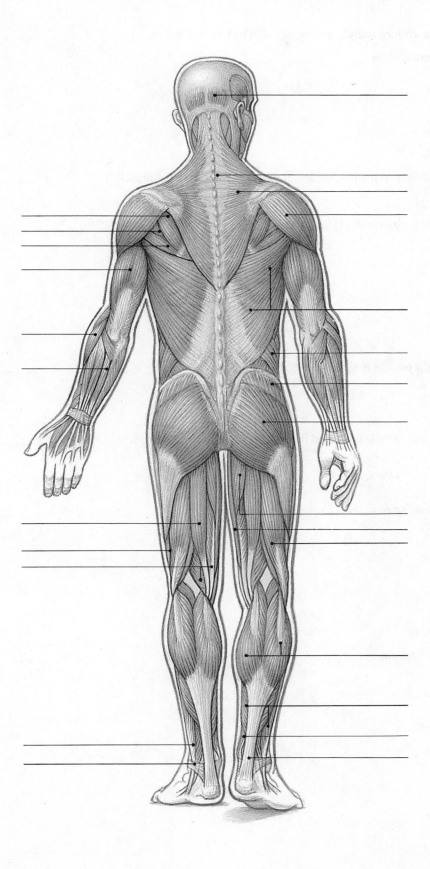

After you have completed the following activities, compare your answers with those given at the end of the manual.

1 The two ends of a skeletal muscle are usually attached by tendons to:

   a mucous membranes.

   b bones.

   c fascia.

   d other muscles.

   e all of the above.

2 The muscle attachment to the bone of lesser movement is called the muscle's

   _____.

3 Which of the following muscles is *not* found in the facial muscle group?

   a risorius

   b corrugator

   c frontalis

   d buccinator

   e triangularis

4 The brachioradialis muscle is located in the forearm.

   a true

   b false

5 The coracobrachialis muscle is located in the leg.

   a true

   b false

6 Which of the following muscles is important in speech?

   a stylohyoid

   b digastric

   c corrugator

   d risorius

   e none of the above

7 The muscle that closes the eye is the:

   a buccinator.

   b frontalis.

   c orbicularis oculi.

   d orbicularis oris.

   e temporalis.

8 When making a fist, what muscle do you use?   _____

    a flexor digitorium superficialis.

    b extensor carpi ulnaris.

    c biceps

    d extensor digitorium communis

    e medial pterygoid

9 The triceps muscle radially rotates the arm.   _____

    a true

    b false

10 Which of the following is the "hamstring" muscle?   _____

    a semitendinosus

    b adductor magnus

    c rectus femoris

    d vastus lateralis

    e transversus

11 Match the following muscles with their action or location:

a ___ frontalis	[1] Puckering muscle.	_____
b ___ auricularis	[2] Elevates lip	_____
c ___ obicularis oris	[3] Covers forehead.	_____
d ___ levator labii superioris	[4] Elevates mandible.	_____
e ___ masseter	[5] Surrounds ear.	_____
f ___ scalenes	[1] Lateral neck.	_____
g ___ serratus anterior	[2] Elevates scapula.	_____
h ___ trapezius	[3] Flexes spine.	_____
i ___ sternohyoid	[4] Anterior neck.	_____
j ___ rectus abdominis	[5] Abducts scapula.	_____
k ___ pectoralis major	[1] Extends forearm.	_____
l ___ deltoid	[2] Flexion of humerus.	_____
m ___ triceps	[1] Flexes wrist.	_____
n ___ biceps	[2] Extends elbow.	_____
o ___ palmaris longus	[3] Flexes elbow.	_____

p ___ pronator teres   [1] Flexes fingers.   _____

q ___ flexor digitorium   [2] Extends little finger.   _____

r ___ extensor carpi ulnaris   [3] Lateral forearm.   _____

s ___ extensor digiti minimi   [4] Extends wrist.   _____

t ___ brachioradialis   [5] Pronates forearm.   _____

u ___ vastus lateralis   [1] Posterior thigh.   _____

v ___ semitendinosus   [2] Adducts thigh.   _____

w ___ extensor hallicus longus   [3] Extends big toe.   _____

x ___ rectus femoris   [4] Flexes ankle.   _____

y ___ tibialis anterior   [5] Extends knee.   _____

12 The muscle that produces a smile is the: _____

     a buccinator.

     b triangularis.

     c mentalis.

     d zygomaticus.

     e temporalis.

13 Which of the following is a suprahyoid muscle?   _____

     a sternohyoid

     b thyrohyoid

     c stylohyoid

     d omohyoid

     e deltoid

14 Match the following muscles with their origins:

     a ___ biceps brachii   [1] Lateral epicondyle of humerus.   _____

     b ___ biceps femoris   [2] Spine of scapula.   _____

     c ___ brachialis   [3] Anterior humerus.   _____

     d ___ deltoid   [4] Ischial tuberosity.   _____

     e ___ extensor carpi radialis   [5] Tubercle of scapula.   _____

     f ___ gracilis   [1] Condyles of femur.   _____

     g ___ gastrocnemius   [2] Superior nuchal line; C_7 to T_{12}.   _____

     h ___ pectoralis major   [3] Body of pubis.   _____

     i ___ rectus femoris   [4] Clavicle, sternum.   _____

     j ___ trapezius   [5] Anterior inferior iliac spine.   _____

k ___ masseter   [1] Linea aspera of femur.   _____

l ___ sternocleidomastoid   [2] Zygomatic arch.   _____

m ___ vastus lateralis   [3] Humerus.   _____

n ___ semimembranosus   [4] Sternum and clavicle.   _____

o ___ triceps   [5] Ischial tuberosity.   _____

15 The trapezius muscle is located in the back region.   _____

    a true

    b false

16 Muscles that assist the agonist are called _____ .   _____

17 The fulcrum lies between the effort and resistance in a _____ - class   _____

    lever system.

18 In a lever system between bones and muscles, the resistance is the effort that is   _____

    overcome by the applied force.

    a true

    b false

19 When a skeletal muscle contracts, a bone always moves.   _____

    a true

    b false

20 Match the following muscles with their insertions:

    a ___ gracilis   [1] Humerus.   _____

    b ___ rectus femoris   [2] Olecranon process.   _____

    c ___ triceps   [3] Patella.   _____

    d ___ gastrocnemius   [4] Calcaneous.   _____

    e ___ deltoid   [5] Tibia.   _____

21 Match the following muscles with their location or action:

    a ___ mentalis   [1] Adduction of humerus.   _____

    b ___ buccinator   [2] Cheek.   _____

    c ___ mylohyoid   [3] Superficial neck.   _____

    d ___ platysma   [4] Upper arm.   _____

    e ___ latissimus dorsi   [5] Lower leg.   _____

    f ___ brachialis   [6] Flexes knee.   _____

    g ___ extensor digitorum longus   [7] Chin.   _____

    h ___ gastrocneumius   [8] Floor of the mouth.   _____

22 Match the following muscles with their location or action:

a ___ adductor magnus   [1] Medial thigh.   _____

b ___ gracilis   [2] Posterior thigh.   _____

c ___ piriformis   [3] Adducts thigh.   _____

d ___ semitendinosus   [4] Posterior pelvis.   _____

e ___ rectus femoris   [5] Anterior thigh.   _____

23 The triceps muscle is an antagonist to the: _____

a brachialis.

b deltoid.

c extensor radialis.

d flexor radialis.

e none of the above.

24 The muscles of the foot give the body a stable platform..   _____

a true

b false

25 The peroneus tertius muscle flexes and everts the leg.   _____

a true

b false

Nervous Tissue

12

Prefixes

af-	to
astr-	a star
con-	with; together
de-	down from; undoing
dendr-	tree; branched structure
ef-	out
gangli-	swelling
gli-	glue
inter-	between
karyo-	nucleus
nerve-	nerve
oligo-	few
peri-	around
pol-	axis; having poles
post-	after
pre-	before
sub-	under
syn-	together
telo-	an end
vesic-	bladder; blister

Suffixes

-aps	fit; fasten
-cyte	hollow vessel
-fibral	fiber
-lemm	rind or peel

DEVELOPING YOUR OUTLINE

Read Chapter 12 in the textbook, focusing on the major objectives. As you progress through it, with your textbook open, answer and/or complete the following When you complete this exercise, you will have a thorough outline of the chapter.

I. Introduction (p. 293)

1 What two organ systems are specialized to maintain homeostasis?

a

b

2 Chemical messengers that communicate between nerve cells are called

_____ .

3 The central nervous system consists of the _____ and

_____ .

OBJECTIVE 1

4 The peripheral nervous system consists of _____ .

II. Neurons: Functional Units of the Nervous System (p. 293)

1 What are the two important properties of neurons?

a

b

A Parts of a Neuron (p. 293)

1 In the space below, draw a diagram of a typical neuron. Label the following: cell body (nucleus, nucleolus, Nissl bodies, ER, mitochondria, neurofilaments, neurotubules, Golgi apparatus), dendrites, and axon (axon hillock, initial segment, synaptic boutons).

2 Complete the following table:

Nerve cell organelle/structure	Function
chromatophilic substance (Nissl bodies)	**a** _____
Golgi apparatus	**b** _____
neurotubules	**c** _____
neurofilaments	**d** _____
dendrites	**e** _____
axon	**f** _____
axoplasm	**g** _____
axolemma	**h** _____
axon hillock	**i** _____
initial segment	**j** _____
collateral branches	**k** _____
telodendria	**l** _____
synaptic bouton	**m** _____
synapse	**n** _____

Nerve cell organelle/structure	Function
myelin/myelin sheath	o _____
neurolemmocytes (Schwann cells)	p _____
neurilemma	q _____
neurofibril nodes (nodes of Ranvier)	r _____
internode	s _____
satellite	t _____

OBJECTIVE 5

3 What three factors affect nerve-impulse conduction?

a

b

c

OBJECTIVE 6

B Types of Neurons: Based on Function (p. 296)

1 Describe each of the following types of neurons in the PNS:

a afferent

b efferent

c mixed

d interneurons

e relay neurons

f local circuit neurons

2 Which of the above six neurons are the three types of peripheral neurons? _____, _____, _____

C Types of Neurons: Based on Structure (p. 297)

    1 Identify the following schematic neurons based on structure:

    2 Identify and describe the following functional segments of a schematic neuron.:

III. Physiology of Neurons (p. 298)

    A Characteristics of the Neuron Plasma Membrane (p. 299)

        1 What channels and pumps are in the nerve membrane?

    B Resting Membrane Potential (p. 299)

        1 Place the following in the proper place on the schematic of a resting nerve cell:

Na^+ (145 mEq/L), Na^+ (12 mEq/L), K^+ (4 mEq/L), K^+ (155 mEq/L), Cl^- (120 mEq/L), Cl^- (4 mEq/L), $Protein^-$ (155 mEq/L).

Outside axon

Inside axon

Now add up the numbers. What is the total positive charge on the outside and inside of the axon?

Outside = _____ .   Inside = _____ .

C The Mechanism of Nerve Impulses (p. 299)

OBJECTIVE 8

1 Fill in the blanks in the following paragraph on the mechanism of a nerve action potential.

A stimulus strong enough to initiate an impulse is called a

a _____. When there is a reversal of ion charges on the

membrane, this is called **b** _____ . When the stimulus is

strong enough to cause depolarization, the neuron **c** _____.

Once a patch on an axon is depolarized, an **d** _____ is

initiated. During the **e** _____ the resting potential of the fiber

membrane is being restored at the part of the membrane where the

impulse has just passed. In other words, the membrane is

f _____ and ready to receive another nerve impulse.

2 Describe the all-or-none law.

D Synapses (p. 300)

OBJECTIVE 9

1 What is a synapse?

2 Describe the following with respect to synapses:

OBJECTIVE 10

a presynaptic neuron

b postsynaptic neuron

c axodendritic synapse

d axosomatic synapse

e axoaxonic synapse

f electrical synapse

g chemical synapse

h neurotransmitter

i synaptic cleft

3 Complete the following table on neurotransmitters:

Group	Transmitter	Function
acetylcholine	a _____	Neuromuscular transmission.
monoamine	b _____	Modulates activity of adrenergic neurons.
monoamine	c _____	Involved in sleep.
d _____	GABA	Evokes IPSPs in brain neurons.
e _____	somatostatin	Inhibits secretion of growth hormone.

E Degeneration and Regeneration of Nerve Fibers (p. 303)

    1 When are nerve fibers able to regenerate?

    2 Describe what happens during nerve regeneration.

F Regeneration in the Central Nervous System (p. 303)

    1 What are three reasons damaged neurons are unsuccessful at regenerating?

    a

    b

    c

IV. Associated Cells of the Nervous System (p. 305)

    A Neuroglia: Associated Cells of the Central Nervous System (p. 305)

        1 Complete the table on neuroglia cells of the CNS.

Type	Description	Function
astrocytes	a _____	b _____
c _____	few short processes	Support framework.
microglia	d _____	Phagocytic.
e _____	single layer, elongated cells	Form inner part of neural tube.

    B Associated Cells of the Peripheral Nervous System (p. 305)

        1 What is a satellite cell?

2 What is the function of neurolemmocytes (Schwann cells)?

V. Organization of the Nervous System (p. 307)

  A Central Nervous System (p. 307)

    1 This system consists of the _____ and _____ .

    2 The major function of the CNS is to act as a _____ system.

  B Peripheral Nervous System (p. 307)

    1 What are the three component parts of the PNS?

      a

      b

      c

    2 What are the two types of nerve cells present in the PNS?

      a

      b

    3 On a purely functional basis, the PNS can be divided into the _____ and _____ systems. Each of these systems is composed of an afferent or _____ and efferent or _____ division.

    4 Complete the following table describing the functions of the different parts of the nervous system:

Division	Function(s)
somatic afferent	**a** _____
somatic efferent	**b** _____
visceral afferent	**c** _____
visceral efferent	**d** _____
sympathetic	**e** _____
parasympathetic	**f** _____

VI. Neuronal Circuits (p. 310)

    1 What three types of neurons are involved in a neuronal circuit? What role does each type play?

      a

      b

      c

A Divergence and Convergence (p. 310)

    1 Sketch the following circuits:

      a divergence

      b convergence

    2 Which of the above two circuits is an example of spatial summation?

B Feedback Circuit (p. 310)

    1 Sketch a negative feedback circuit.

C Parallel Circuits (p. 311)

    1 Sketch a parallel circuit.

D Two-Neuron Circuit (p. 311)

    1 Describe this type of circuit.

E Three-Neuron Circuit (p. 311)

    1 Sketch a three-neuron circuit.

VII. The Effects of Aging on the Nervous System (p. 312)

    1 How does aging affect the nervous system?

    2 What are some common ailments of aging?

VIII. When Things Go Wrong (p. 312)

A Describe the following diseases:

    1 multiple sclerosis

    2 amyotrophic lateral sclerosis

OBJECTIVE 14

OBJECTIVE 15

3 peripheral neuritis

4 myasthenia gravis

5 Parkinson's disease

6 Huntington's chorea

These are terms you should know before proceeding to the post-test.   **MAJOR TERMS**

stimuli *(p. 293)*

integration *(p. 293)*

central nervous system (CNS) *(p. 293)*

peripheral nervous system (PNS) *(p. 293)*

neurons *(p. 293)*

cell body *(p. 293)*

dendrites *(p. 293)*

axon *(p. 295)*

end bulbs (synaptic boutons) *(p. 295)*

synapse *(p. 295)*

myelin *(p. 295)*

myelin sheath *(p. 295)*

neurolemmocyte (Schwann cell) *(p. 295)*

neurilemma *(p. 295)*

neurofibral nodes (nodes of Ranvier) *(p. 295)*

nerve *(p. 295)*

afferent (sensory) neurons *(p. 296)*

efferent (motor) neurons *(p. 296)*

effectors *(p. 296)*

interneurons *(p. 297)*

multipolar neurons *(p. 297)*

bipolar neurons *(p. 297)*

unipolar neurons *(p. 297)*

receptive segment *(p. 297)*

initial segment *(p. 298)*

conductive segment *(p. 298)*

transmissive segment *(p. 298)*

resting membrane potential *(p. 299)*

sodium-potassium pump *(p. 299)*

threshold stimulus *(p. 299)*

depolarization *(p. 299)*

action potential (nerve impulse) *(p. 299)*

repolarized *(p. 300)*

refractory period *(p. 300)*

all-or-none law *(p. 300)*

saltatory conduction *(p. 300)*

presynaptic neuron *(p. 300)*

postsynaptic neuron *(p. 301)*

electrical synapse *(p. 301)*

chemical synapse *(p. 301)*

neurotransmitter *(p. 301)*

neuroglia *(p. 305)*

ganglia *(p. 305)*

satellite cells *(p. 305)*

neurilemma cells *(p. 305)*

afferent (sensory) nerve cells *(p. 307)*

efferent (motor) nerve cells *(p. 307)*

somatic nervous system *(p. 307)*

visceral nervous system *(p. 307)*

autonomic nervous system *(p. 307)*

general somatic afferent fiber *(p. 310)*

general visceral afferent fiber *(p. 310)*

general somatic efferent fiber *(p. 310)*

general visceral efferent fiber *(p. 310)*

special afferent fiber *(p. 310)*

special visceral efferent fiber *(p. 310)*

neuronal circuits *(p. 310)*

nucleus (neuron pool) *(p. 310)*

divergence *(p. 310)*

convergence *(p. 310*

feedback circuit *(p. 310)*

parallel circuit *(p. 311)*

two-neuron circuit *(p. 311)*

three-neuron circuit *(p. 311)*

multiple sclerosis *(p. 312)*

amyotrophic lateral sclerosis *(p. 313)*

peripheral neuritis *(p. 313)*

myasthenia gravis *(p. 313)*

Parkinson's disease *(p. 313)*

Huntington's chorea *(p. 313)*

Based on the terminology contained within this chapter, label the following nerve: **LABELING ACTIVITY**

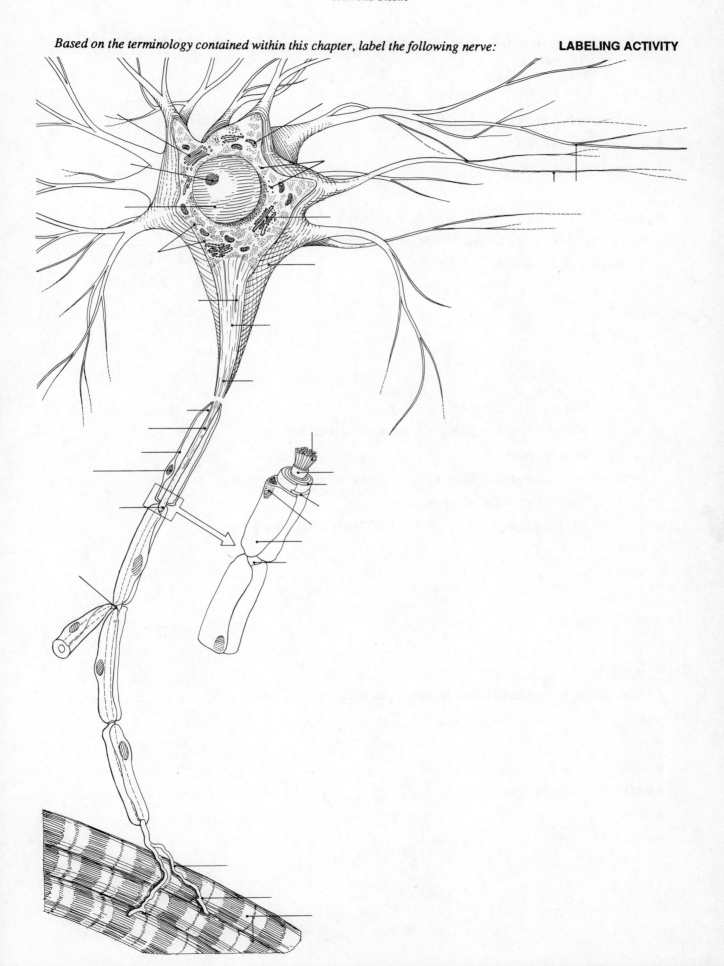

After you have completed the following activities, compare your answers with those given at the end of this manual.

1 Glial cells:

   a act as barriers to chemicals.

   b support neurons physically.

   c sustain neurons metabolically.

   d **a and b.**

   e **a, b,** and **c.**

2 A neuron that has one long axon and multiple, short, highly branched dendrites extending form the cell body is:

   a bipolar.

   b unipolar.

   c multipolar.

   d somatic.

   e visceral.

3 When an inhibitory synapse on cell A fires, it may:

   a cause cell A to decrease the frequency of firing of action potentials if it is already firing action potentials.

   b cause cell A to require greater excitatory input in order to fire action potentials if it is not already firing action potentials.

   c cause cell A to cease firing action potentials if it is already firing action potentials.

   d **a and b.**

   e **a, b,** and **c.**

4 Myelin is a protein.

   a true

   b false

5 Saltatory conduction in nonmyelinated neurons is greater than in myelinated neurons.

   a true

   b false

6 Interneurons lie outside of the CNS.

   a true

   b false

7 Matching.

a ___ receptor site

b ___ action potential

c ___ decremental

d ___ absolute refactory period

e ___ excitatory

f ___ a general potential

g ___ excitability

h ___ synaptic membrane

i ___ neurofibral nodes
(nodes of Ranvier)

j ___ presynaptic neuron

[1] A type of synapse.

[2] Under the synaptic knob.

[3] Conducts information toward a
synapse

[4] Varies with stimulus strength.

[5] Determines whether the synapse

[6] The form in which information
is relayed along neurons.

[7] Intervals along axons.

[8] Amplitude decreasing with
increasing distance.

[9] The time delay between
successive action potentials.

[10] The property that enables neurons
to produce action potentials.

8 Matching.

a ___ myelin

b ___ bipolar

c ___ receptor potential

d ___ presynaptic neuron

e ___ IPSP

f ___ neurotransmitter

g ___ axon

h ___ recruitment

i ___ interneurons

j ___ neurofibral nodes
(nodes of Ranvier)

[1] Extends from a nerve cell body.

[2] Fatty material.

[3] Responsible for saltatory conduction.

[4] Have only two processes.

[5] Account for 99 percent of all
nerve cells.

[6] A generatory potential.

[7] The "calling in" of receptors on
additional nerv cells.

[8] Conducts information across a
synapse.

[9] Transmits information across a
synapse.

[10] Hyperpolarization.

9 Which of the following is *not* a neuromodulator?   _____

   a a histamine

   b a prostaglandin

   c an enkephalin

   d an endorphin

   e a catecholamine

10 Which of the following ions shows *no* net diffusion across a resting nerve   _____

   membrane?

   a sodium

   b potassium

   c chloride

   d a and b

   e a, b, and c

11 The peripheral nervous system may be divided on a functional basis into the   _____

   _____ and _____ nervous systems.   _____

12 The basic properties of nerve cells include _____ and _____. _____

                                                                                   _____

13 Neurons and their peripheral fibers may be classified into functional components as: _____

   a   _____

   b   _____

   c   _____

   d   _____

   e

14 Neurons may be classified according to their structure as: _____

   a   _____

   b   _____

   c

15 Most neurons are composed of specific functional segments known as: _____

   a   _____

   b   _____

   c   _____

   d

16 A resting cell that is not conducting an impulse is polarized.   _____

   a true

   b false

17 The phenomenon of a neuron firing at full power or not at all is the
_____ law.   _____

18 _____ conduction occurs when a nerve impulse jumps from node to   _____
node on a myelinated fiber.

19 List two diseases that are related to improper secretions of neurotransmitters.   _____

a   _____

b

20 Neuronal circuits may be: _____

a   _____

b   _____

c   _____

d   _____

e   _____

f

21 The depolarization of the nerve membrane is produced by any factor that: _____

a increases the membrane's permeability to sodium ions.

b inhibits the membrane's permeability to sodium ions.

c starts the sodium pump.

d increases the membrane's permeability to potassium ions.

e all of the above.

22 The intensity of a stimulus to a nerve bundle is conveyed to the CNS by: _____

a spatial summation.

b temporal summation.

c the frequency of impulses.

d the number of nerve fibers carrying impulses.

e all of the above.

23 All of the following are neurotransmitters *except*:   _____

a GABA.

b acetylcholine.

c epinephrine.

d norepinephrine.

e glutamic acid.

24 Where are the neurotransmitters stored that are secreted into the synaptic cleft   _____
separating two adjacent neurons?

   a in vesicles in the postsynaptic membrane

   b in small vesicles in the synaptic knob of the axon

   c at specific receptor sites on the axon membrane

   d in the synaptic cleft

   e c and **d**

25 What would you expect to occur if acetylcholine was released at a synapse but no   _____
acetylcholinesterase was present?

   a Rapid firing of the postsynaptic neuron would occur.

   b The acetylcholine would fail to diffuse across the synapse

   c The acetylcholine would not combine with the receptor site on the postsynaptic
membrane.

   d A single nerve impulse would be generated.

   e All of the above.

The Spinal Cord and Spinal Nerves

13

Prefixes

arachn-	spider; spider web
corp-	a body
dys-	bad; disordered
ependym-	tunic
epi-	above
funi-	small cord or fiber
gangli-	swelling
lumbo-	loins
menin-	a membrane
plex-	a network
ram-	a branch
recept-	a receiver
sub-	below

Suffixes

-lemm	rind or peel
-sacral	holy

Read Chapter 13 in the textbook, focusing on the major objectives. As you progress through it, with your textbook open, answer and/or complete the following. When you complete this exercise, you will have a thorough outline of the chapter.

DEVELOPING YOUR OUTLINE

I. Introduction (p. 318)

    1 What are the two important functions of the spinal cord?

      a

      b

II. Basic Anatomy of the Spinal Cord (p. 318)

OBJECTIVE 1

    1 Name the two prominent spinal cord enlargements.

      a

      b

2 Of the 31 pairs of spinal nerves and roots, there are _____ cervical pairs, _____ thoracic, _____ lumbar, _____ sacral, and _____ coccygeal.

3 What does *cauda equina* mean in English?

How does this meaning relate to the spinal cord? _____

4 What four specific structures protect the spinal cord and its roots?

a

b

c

d

A Spinal Meninges (p. 318)

1 The three meninges from the outer to inner layer are the:

a

b

c

2 Describe the contents in each of the following spaces:

a epidural

b subdural

c subarachnoid

3 A saddle block anesthetic is given into which meningeal space?

B Cerebrospinal Fluid (p. 318)

    1 List two mechanisms involved in CSF production.

        a

        b

    2 Describe CSF.

    3 What are some functions of CSF?

    4 What are some clinical uses of CSF?

    5 a Where, specifically, is a lumbar puncture performed?

        b Where, specifically, is a cistern puncture performed?

C Internal Structure (p. 321)

OBJECTIVE 2

    1 If you cut a spinal cord in cross section, what structures would be visible?

    2 Specifically, what is the gray matter of the spinal cord composed of?

    3 Give the functions of the following spinal cord structures:

        a posterior horns

        b anterior horns

        c gray commissure

        d ventral roots

        e dorsal roots

        f funiculi

        g fasciculi

h ascending tracts

i descending tracts

III. Functional Roles of Pathways of the Central Nervous System (p. 324)

OBJECTIVE 3

A General Somatic Efferent (Motor System) (p. 324)

    1 What are the two types of lower motor neurons?

      a

      b

    2 What are the three upper motor neuron pathways?

      a

      b

      c

B Processing Centers (p. 327)

    1 What are the three parts to each processing center?

      a

      b

      c

C Sensory Pathways (p. 327)

    1 Describe each of the following:

      a first-order neuron

      b second-order neuron

      c third-order neuron

    2 What happens when sensory neurons decussate?

    3 How are some spinal tracts named?

D Anterolateral System (p. 328)

1 Describe the composition of the anterolateral system.

2 Describe the composition of the lateral spinothalamic tract.

3 Describe the spinoreticulothalamic pathway.

4 What is the functional significance of decussation?

E Posterior Column-Medial Lemniscus Pathway (p. 328)

1 Describe the posterior column-medial lemniscus pathway.

IV. Spinal Reflexes (p. 329)

OBJECTIVE 4

1 What is the difference between a somatic and visceral reflex?

2 Describe a:

a monosynaptic reflex arc

b polysynaptic reflex arc

3 What is the difference between a spinal reflex and a simple reflex?

4 Complete the following descriptions of the stretch reflex:

a This reflex is also known commonly as the _____ reflex.

b Since the same side of the body is involved, this can be called a _____ reflex.

c The stretch receptors in the quadricep muscle that are stimulated are called _____ .

d Nerve impulses are conveyed to the L2 or L3 levels of the spinal cord via _____ fibers.

e In the spinal cord, afferent neurons synapse with lower motor neurons called _____

5 Fill in the blanks in the following paragraph on the gamma motor neuron reflex arc:

OBJECTIVE 5

a Thinly myelinated fibers called _____ innervate tiny muscles in a muscle spindle called _____.

These are in contrast to the _____ muscle fibers that do the work of contraction.

6 The two main functions of the gamma motor neuron reflex arc are:

a

b

7 List, in correct sequence, the components of the polysynaptic reflex arc.

8 Another term for the sensory receptors in the above schematic is

_____ .

9 According to the principle of _____, the contraction of the

_____ muscles is synchronized with the relaxation of the

_____ muscles.

10 What is the importance of the crossed extensor reflex?

A Functional and Clinical Aspects of Reflex Responses (p. 331)

1 The continuous state of muscle contraction is known as muscle

_____ .

2 When a muscle loses some of its tonus, this is called _____ .

3 The loss of all of a muscle's tonus is called _____ .

4 The increased tonus of a muscle is called _____ .

5 A condition in which a reflex is less responsive than normal is

_____ . A condition in which a reflex is more responsive than

normal is _____ .

6 Complete the following table on diagnostic reflexes of the central nervous system:

Reflex	Description	Indication
plantar (Achilles)	**a** _____	Lesions of peripheral nerves.
Babinski's	plantar reflex	**b** _____
c _____	flexion of neck; flexes legs	Irritation of meninges.
patellar	**d** _____	Damage at L2–L4 levels.
Romberg's	**e** _____	Indicates dorsal column injury.

V. Structure and Distribution of Spinal Nerves (p. 332)

OBJECTIVE 7

A How Spinal Nerves Are Named (p. 332)

    1 How are spinal nerves named?

OBJECTIVE 8

B Structure of Spinal Nerves (p. 333)

    1 What are the three sheaths of connective tissue around spinal nerve fibers?

        a

        b

        c

OBJECTIVE 9

C Branches of Spinal Nerves (p. 333)

    1 Complete this table on the branches of a spinal nerve.

Branch name	Innervation
a _____	skin of the back
ventral ramus	**b** _____
c _____	vertebrae, meninges
d _____	visceral structures

D Plexuses (p. 333)

    1 Summarize the plexuses by completing this table.

Name	Spinal nerves involved	Location
a _____	C1–C4	neck
brachial	b _____	lower neck
lumbar	L1–L4	c _____
d _____	L4, L5, S1, S3	e _____
coccygeal	coccygeal nerve plus S4 and S5	f _____

    2 The ventral rami of nerves _____ through _____

    do not form plexuses.

    3 The cervical plexus nerves can be placed into four groups. What are

    they, and what does each innervate?

      a

      b

      c

      d

    4 List the five major nerves of the brachial plexus and describe what each
    one innervates.

      a

      b

      c

      d

      e

5 List the two major nerves of the lumbar plexus.

a

b

6 List the four major nerves of the sacral plexus.

a

b

c

d

7 What three nerves form the coccygeal plexus?

a

b

c

E Intercostal Nerves (p. 335)

OBJECTIVE 10

1 The intercostal nerves are _____ through _____ spinal nerves.

2 What is the function of the intercostal nerves?

F Dermatomes (p. 339)

OBJECTIVE 11

1 What is a dermatome?

2 How many dermatomes are there?

VI. Developmental Anatomy of the Spinal Cord (p. 339)

1 What is a notocord?

2 What is a neurotube?

3 How does the spinal cord develop?

VII. When Things Go Wrong (p. 341)

A Spinal Cord Injury (p. 341)

    1 Complete the following table on spinal cord injuries:

Terminology (name)	Description
a _____	Any lesion of the spinal cord that damages its neurons.
b _____	Severing the spinal cord.
areflexia	c _____
d _____	Loss of sensation below the level of injury.
e _____	Motor loss in both lower extremities.
quadriplegia	f _____
g _____	Paralysis of upper and lower limbs on one side of the body.

B Carpal Tunnel Syndrome (p. 342)
    1 What causes carpal tunnel syndrome?

C Poliomyelitis (p. 342)
    1 The virus that causes polio has an affinity for what part of the nervous system? What effect does this virus have?

D Sciatica (p. 343)
    1 What is sciatica?

    2 What is the most common cause of sciatica?

E Shingles (p. 343)
    1 What part of the nervous system is involved in shingles?

F Spinal Meningitis (p. 343)
    1 What are four initial signs of meningitis?

        a

        b

c

d

2 What effects does meningitis have on the nervous system?

These are terms you should know before proceeding to the post-test.   **MAJOR TERMS**

spinal cord *(p. 318)*

cervical enlargement *(p. 318)*

lumbosacral enlargement *(p. 318)*

cauda equina *(p. 318)*

meninges *(p. 318)*

dura mater *(p. 318)*

arachnoid *(p. 318)*

pia mater *(p. 318)*

epidural space *(p. 318)*

subdural space *(p. 318)*

subarachnoid space *(p. 318)*

cerebrospinal fluid (CSF) *(p. 318)*

gray matter *(p. 321)*

posterior horns *(p. 321)*

anterior horns *(p. 321*

lateral horns *(p. 321)*

gray commissure *(p. 321)*

white matter *(p. 321)*

funiculi *(p. 321)*

tracts *(p. 321)*

ascending tracts *(p. 321)*

descending tracts *(p. 321)*

ventral roots *(p. 324)*

dorsal roots *(p. 324)*

dorsal root ganglia *(p. 324)*

processing centers *(p. 324*

upper motor neuron *(p. 324)*

lower motor neuron *(p. 324)*

neuromuscular spindles *(p. 326)*

first-order neuron *(p. 327)*

second-order neuron *(p. 327)*

third-order neuron *(p. 327)*

decussate *(p. 328)*

anterolateral system *(p. 328)*

lateral spinothalamic tract *(p. 328)*

spinoreticulothalamic pathway *(p. 328)*

anterior spinothalamic tract *(p. 328)*

posterior column-medial lemniscus pathway *(p. 328)*

reflex *(p. 329)*

reflex arc *(p. 329)*

stretch (myotatic) *(p. 329)*

gamma motor neuron reflex arc *(p. 329)*

withdrawal reflexes *(p. 330)*

muscle tone *(p. 331)*

hypotonia *(p. 331)*

atonia *(p. 331)*

hypertonia *(p. 331)*

hyporeflexia *(p. 331)*

hyperreflexia *(p. 331)*

dorsal root *(p. 332)*

mixed nerve *(p. 332)*

fascicles *(p. 333)*

epineurium *(p. 333)*

perineurium *(p. 333)*

endoneurium *(p. 333)*

rami *(p. 333)*

dorsal ramus *(p. 333)*

ventral ramus *(p. 333)*

meningeal ramus *(p. 333)*

rami communicantes *(p. 333)*

plexuses *(p. 333)*

cervical plexus *(p. 333)*

brachial plexus *(p. 335)*

lumbar plexus *(p. 335)*

sacral plexus *(p. 335)*

coccygeal plexus *(p. 335)*

intercostal nerves *(p. 335)*

dermatome *(p. 339)*

neural tube *(p. 339)*

neural crest *(p. 339)*

spinal cord injury *(p. 341)*

paraplegia *(p. 342)*

quadriplegia *(p. 342)*

hemiplegia *(p. 342)*

carpal tunnel syndrome *(p. 342)*

poliomyelitis *(p. 342)*

sciatica *(p. 342)*

shingles *(p. 342)*

spinal meningitis *(p. 342)*

REGISTERED NURSE

Registered nurses (RNs) perform a wide variety of functions. They observe patients and asses and record symptoms, reactions, and progress of patients. Registered nurses are trained to administer medication, assist in the rehabilitation of patients, instruct patients and family members in proper health maintenance care, and help maintain a physical and emotional environment that promotes recovery. Some registered nurses provide nursing services in hospitals and nursing homes, Others are active in research or instruct students. The scope of the nurse's responsibilities is determined by the setting. Hospital nurses, private duty nurses, and nurse educators all have varying duties. All can be RNs.

A license is required to practice professional nursing. To obtain a license, a nurse must be a graduate of a state-approved school of nursing and must pass a written state board competence examination. Three types of educational programs—diploma, bachelor's degree, and associate degree—prepare candidates for licensure.

Based on the terminology contained within the chapter, label the following cross section of a spinal cord:

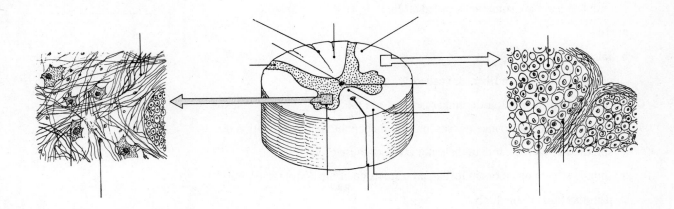

After you have completed the following activities, compare your answers with those given at the end of this manual.

1 In the meninges, the large blood vessels pass through the:

   a dura mater.

   b pia mater.

   c subdural space.

   d subarachnoid space.

   e epidural space.

2 Cerebrospinal fluid is produced and secreted by the:

   a meninges.

   b choroid plexus.

   c ventricles.

   d blood.

   e pia mater.

3 The white matter of the spinal cord is made up of:

   a myelinated axons.

   b unmyelinated axons.

   c cell bodies of neurons.

   d axons.

   e all of the above.

4 A person receives a bullet wound severing the spinal nerves at the S1 and S2 levels. Where would motor function be lost?

_____

_____

_____

5 At what level is a spinal tap performed? _____

6 Afferent neurons are sometimes called afferents or first-order neurons. _____

   a true

   b false

7 Bundles of spinal nerve fibers are called _____. _____

8 The second, third, and fourth lumbar nerves form the _____ plexus. _____

9 Alpha motor neurons to an extensor muscle can be inhibited by activation of: _____

   a spindle stretch receptors in the same extensor muscle.

   b spindle stretch receptors in ipsilateral (same side of the body) flexor muscles.

   c pain receptors in the limb.

   d gamma efferents in the extensor muscle.

   e both **b** and **c**.

10 If a person steps on a tack with his or her right foot, which of the following will occur? _____

   a The foot will begin to withdraw from the tack before the brain is conscious of
the injury (feels pain).

   b Flexor muscles in the right leg and extensor muscles in the left leg will be
activated.

   c Extensor motor neurons in the right leg will be inhibited by a monosynaptic
reflex arc from receptors in the injured area.

   d **a** and **b**.

   e **a**, **b**, and **c**.

11 Which of the following is characteristic of the knee-jerk reflex? _____

   a A quick stretch of the muscle spindle results in contraction of the extrafusal
muscle fibers.

   b Increased stimulation of the gamma fibers decreases the response of the
extrafusal muscle fibers to a quick stretch.

   c Damage to the afferent pathway results in a complete loss of voluntary contrac
tion of the extrafusal fibers.

   d Contraction of the extrafusal fibers decreases the number of action potentials
per unit time in the afferent pathway from the muscle spindle.

   e **a** and **b**.

12 Which of the following structures is *not* included in the gamma loop?   _____

   a tendon organ

   b gamma motor neurons

   c alpha motor neurons

   d muscle spindles

   e involuntary muscles

13 The fibers of the direct corticospinal pathway end on the: _____

   a interneurons.

   b alpha motor neurons.

   c gamma motor neurons.

   d a and b.

   e a, b, and c.

14 Which of the following are components of the reflex arc?   _____

   a receptor

   b afferent pathway

   c integrating center

   d efferent pathway

   e all of the above.

15 Matching.

   a ___ basal ganglia   [1] Knee jerk.   _____

   b ___ monosynaptic   [2] Monitor tension.   _____

   c ___ stretch reflex   [3] Involved with very rapid movements.   _____

   d ___ reciprocal innervation [4] Located on top of the brainstem.   _____

   e ___ cerebellum   [5] Involved with slow, continuous   _____
                                    movements.

   f ___ tendon organ   [6] Skeletal muscle cells outside a   _____
                                    muscle spindle.

   g ___ cerebrum   [7] Without the interposition of any   _____
                                    interneurons.

   h ___ intention tremor   [8] The excitation of one muscle and   _____
                                    the simultaneous inhibition of its
                                    antagonistic muscle.

   i ___ corticospinal pathway [9] Also called the pyramidal tract.   _____

   j ___ extrafusal   [10] An oscillating tremor.   _____

16 Matching.

a ___ alpha   [1] These are actions we think about   _____

b ___ extrafusal   [2] Skeletal muscle cells outside the   _____
                                      muscle spindles.

c ___ gamma   [3] Motor neurons that control   _____
                                      skeletal fibers.

d ___ extrapyramidal system   [4] Neurons innervating spindle fibers.   _____

e ___ gravity   [5] When alpha and gamma neurons   _____
                                      fire together.

f ___ substantia nigra   [6] Multineuronal pathway other than   _____
                                      pyramidal.

g ___ athetosis   [7] Posterior part of the frontal lobe   _____
                                      of the cerebrum.

h ___ motor cortex   [8] Hyperextension of the fingers.   _____

i ___ voluntary   [9] A necessary component of   _____
                                      locomotion.

j ___ coactivated   [10] A subcortical nucleus.   _____

17 The plexus from which the sciatic nerve arises is the: _____

a sacral.

b brachial.

c cervical.

d lumbar.

e none of the above.

18 The spinal tract carrying voluntary motor impulses to skeletal muscles is the: _____

a gracilis.

b corticospinal.

c rubrospinal.

d spinocerebellar.

e red nucleus.

19 A temporary synaptic depression after spinal trauma is called _____.   _____

20 There are _____ pairs of cervical spinal nerves.   _____

21 The adult spinal cord extends the length of the spinal column.   _____

a true

b false

22 _____ is a form of nerve inflammation characterized by sharp pains along the sciatic nerve and its branches.

_____

23 The motor or sensory loss of function in both lower extremities is called _____ .

_____

24 Several _____ fascicles are held together by a sheath called the _____; the _____ encases each fascicle and each nerve fiber is covered by the _____ .

_____

_____

_____

_____

25 The motor pathways and associated neuronal circuits are classified as _____ or _____ systems.

_____

_____

The Brain and Cranial Nerves

14

Prefixes

arachn-	spider
cephalo-	head
cerebro-	brain
cort-	shell; bark
hemi-	half
medull-	under; beneath
peri-	around
pon-	a bridge
tom-	cutting; a segment
top-	place; region

Suffixes

-duct	from; to draw
-gram	wrtten; drawn
-graph	write; draw

Read Chapter 14 in the textbook, focusing on the major objectives. As you progress through it with your textbook open, answer and/or complete the following. When you complete this exercise, you will have a thorough outline of the chapter.

DEVELOPING YOUR OUTLINE

I. Introduction (p. 348)

1 There are 100 billion neurons in the brain. What other structure is composed of 100 billion units?

II. General Structure of the Brain (p. 348)

OBJECTIVE 1

1 Technically, the brain should be called the _____ .

2 The four major division of the brain are:

a

b

c

d

3 What subdivisions comprise the diencephalon?

a

b

c

d

4 What subdivisions comprise the brainstem?

a

b

c

5 Where is the cerebellum located?

6 What structure connects the two cerebral hemispheres?

III. Meninges, Ventricles, and Cerebrospinal Fluid (p. 348)

1 Name four ways the brain is protected and supported.

a

b

c

d

A Cranial Meninges (p. 348)

1 List the three cranial meninges.

a

b

c

2 List three ways the cerebral meninges differ from the spinal meninges.

a

b

c

3 In what space is CSF located?

4 Describe the location of the pia mater.

B Ventricles of the Brain (p. 350)

    1 List in order, from the top of the brain downward, the ventricles of the brain.

    a

    b

    c

    2 Each lateral ventricle is connected to the third ventricle via the

    _____.

    3 The third ventricle is connected to the fourth ventricle via the

    _____.

C Cerebrospinal Fluid in the Brain (p. 352)

    1 Circle the correct answer(s) about CSF.

    a An adult has about _____ ml of CSF.

      1 10   **2** 20   **3** 50   **4** 125   **5** 200

    b Cerebrospinal fluid is:

      1 clear and colorless.   **2** yellow.   **3** red.   **4** white.

      5 green.

    c Cerebrospinal fluid functions in:

      1 brain buoyancy.

      2 controlling the chemical environment of CNS.

      3 the exchange of nutrients.

      4 transport.

      5 all of the above.

   d Cerebrospinal fluid is formed by:

      1 diffusion.

      2 active transport.

      3 facilitated diffusion.

      4 1 and 2.

      5 1, 2, and 3.

2 Beginning with the two lateral ventricles, name the structures that CSF passes through until it reaches the cisterna magna.

   a two lateral ventricles

   b

   c

   d

   e

   f

   g cisterna magna

IV. Nutrition of the Brain (p. 353)

OBJECTIVE 4

   A Effects of Deprivation (p. 353)

      1 What two substances does the brain need continually?

        a

        b

      2 Why is it imperative that the blood vessels of the brain do not constrict easily?

V. Brainstem (p. 356)

OBJECTIVE 5

      1 What are the three segments of the brainstem?

        a

b

c

2 Overall, what is the function of the brainstem?

A Neural Pathways of Nuclei and Long Tracts (p. 356)

    1 Name the five sensory (ascending) pathways in the brainstem.

        a

        b

        c

        d

        e

    2 Name the four motor (descending) pathways in the brainstem.

        a

        b

        c

        d

OBJECTIVE 6

B Reticular Formation (p. 356)

    1 What are the three major parts of the reticular formation?

        a

        b

        c

2 Describe several functions of the reticular formation.

C Reticular Activating System (p. 357)

    1 Name two nuclear groups located in this system and the neurotransmitter secreted by each.

      a

      b

    2 The RAS is sometimes referred to as the _____ system.

    3 Describe some of the functions of the RAS system.

D Medulla Oblongata (p. 358)

    1 Describe the specific nuclei that are located in the medulla.

    2 What is the pyramidal decussation?

    3 From the brain's perspective, what is the significance of pyramidal decussation?

E Pons (p. 358)

    1 The pons can be divided into what two regions?

      a

      b

    2 What two centers are located within the reticular formation of the dorsal pons?

F Midbrain (p. 361)

    1 Specifically, where is the midbrain located?

    2 What are the corpora quadrigemina?

    3 What are the functions of the following structures?

      a ruber nucleus

      b substantia nigra

VI. Cerebellum (p. 361)

OBJECTIVE 7

   A Anatomy of the Cerebellum (p. 361)

      1 What are the three anatomical parts of the cerebellum?

         a

         b

         c

      2 Describe each of the following anatomical parts of the cerebellum:

         a cerebellar cortex

         b folia cerebelli

         c fissures

         d arbor vitae

         e cerebellar peduncles

   B Functions of the Cerebellum (p. 363)

      1 What are the main functions of the cerebellum?

VII. Cerebrum (p. 363)

OBJECTIVE 8

   A Anatomy of the Cerebrum (p. 363)

      1 Describe each of the following anatomical parts of the cerebrum:

         a cerebral cortex

         b gyri

         c sulci

         d fissures

2 Complete the following table on fibers within the white matter:

Fiber type	Function
association	a _____
b _____	Axons that project to opposite cortical area.
c _____	Project to other areas of the brain outside the cerebral cortex.

3 Describe the function of the basal ganglia.

B Functions of the Cerebrum (p. 367)

  1 The general functions of the left cerebral hemisphere are:

  2 The general functions of the right cerebral hemisphere are:

C Cerebral Lobes (p. 367)

OBJECTIVE 9

OBJECTIVE 10

  1 Complete the following table on the six lobes of the cerebral hemisphere:

Lobe	Functions
frontal	a _____
b _____	evaluation of general senses (sometimes information)
temporal	c _____
d _____	visual sensations
e _____	olfaction, emotions, behavior, memory
1. hippocampus	f _____
2. g _____	fear, fight, or flight
3. hypothalamus	h _____

VIII. Diencephalon (p. 370)

  1 List the four anatomical parts of the diencephalon.

OBJECTIVE 11

    a

    b

c

d

A Thalamus (p. 371)

    1 Describe the four major areas of activity associated with the thalamus.

      a

      b

      c

      d

    2 Where is the hypothalamus located?

B Hypothalamus (p. 372)

    1 Describe the eight most important functions of the hypothalamus.

      a

      b

      c

      d

      e

      f

      g

      h

C Epithalamus (p. 373)

    1 Describe the three parts of the epithalamus.

      a

   b

   c

   2 What general function does the pineal gland have?

D Ventral Thalamus (p. 373)
   1 What important nucleus is located in the ventral thalamus?

IX. Cranial Nerves (p. 373)

OBJECTIVE 12

   1 Answer the following questions about the cranial nerves:

   There are a _____ pairs of cranial nerves. Their names are an

   indication of some **b** _____ or **c** _____ feature of

   the nerve and their **d** _____ numerals indicate the

   e _____ Cranial nerves **f** _____ and

   g _____ are nerves of the cerebrum, and nerves

   h _____ through **i** _____ are nerves of the

   brainstem. Of the ten brainstem nerves, one (number **j** _____)

   is purely a sensory nerve, five are primarily motor nerves (numbers

   k _____, _____, _____

   _____, _____), and four are mixed nerves

   (numbers **l** _____, _____, _____,

   _____). The **m** _____ fibers of the cranial nerves

   emerge from the brainstem, whereas the **n** _____ fibers

   emerge from neurons with cell bodies outside the brain.

Answer the following questions about the respective cranial nerves:

   A Cranial Nerve I: Olfactory (p. 376)

      1 type _____

      2 origin _____

      3 function _____

   B Cranial Nerve II: Optic (p. 376)

      1 type _____

      2 origin _____

      3 cross at the _____

      4 function _____

C Cranial Nerve III: Oculomotor (p. 376)

    1 type _____

    2 origin _____

    3 function _____

D Cranial Nerve IV: Trochlear (p. 377)

    1 type _____

    2 origin _____

    3 function _____

E Cranial Nerve V: Trigeminal (p. 377)

    1 The three branches of this nerve are:

        a

        b

        c

    2 Is this the largest cranial nerve?

F Cranial Nerve VI: Abducens (p. 377)

    1 type _____

    2 origin _____

    3 function _____

G Cranial Nerve VII: Facial (p. 379)

    1 type _____

    2 origin _____

    3 function _____

H Cranial Nerve VIII: Vestibulocochlear (p. 379)

    1 Name the two branches and their types, origins, and functions.

        a

        b

I Cranial Nerve IX: Glossopharyngeal (p. 380)

    1 type _____

    2 origin _____

    3 function _____

J Cranial Nerve X: Vagus (p. 380)

    1 type _____

    2 uniqueness _____

    3 function _____

K Cranial Nerve XI: Accessory (p. 382)

    1 type _____

    2 origin _____

    3 function _____

L Cranial Nerve XII: Hypoglossal (p. 383)

    1 type _____

    2 origin _____

    3 function _____

X. Development of the Brain (p. 383)

    1 Describe how the brain develops.

XI. When Things Go Wrong (p. 385)

OBJECTIVE 13

  A Senility (p. 385)

    1 Severe atrophy of the brain is commonly called _____ .

    2 What are some characteristics of senile dementia?

    3 Is senility a normal condition of aging?

  B Alzheimer's Disease (p. 385)

    1 Describe Alzheimer's disease.

    2 How do senile dementia and Alzheimer's disease differ?

  C Cerebral Palsy (p. 385)

    1 Name the three major types of cerebral palsy.

      a

      b

      c

2 What system does this disease affect?

D Cerebrovascular Accident (CVA) (p. 386)

    1 A CVA is commonly called a _____ .

    2 List and describe four causes of a CVA.

        a

        b

        c

        d

E Epilepsy (p. 386)

    1 Describe the following types of epilepsy:

    a symptomatic

    b idiopathic

    c grand mal

    d petit mal

F Headache (p. 386)

    1 Describe the following types of headaches:

    a migraine

    b tigeminal neuralgia

    c cranial arteritis

G Parkinson's Disease (p. 386)

    1 What is believed to be the cause of Parkinsonism?

H Dyslexia (p. 386)

    1 What is dyslexia?

I Encephalitis (p. 386)

    1 What are some causes of encephalitis?

J Bell's Palsy and Other Disorders of Cranial Nerves (p. 387)

    1 Match the names of the disorder with the correct cranial nerves.

a ___ Bell's palsy	**[1]** Olfactory
b ___ diplopia	**[2]** Optic
c ___ tic douloureux	**[3]** Oculomotor
d ___ ptosis	**[4]** Trochlear
e ___ glaucoma	**[5]** Trigeminal
f ___ anosmia	**[6]** Abducens
g ___ dysarthia	**[7]** Facial
h ___ tinnitus	**[8]** Vestibulocochlear
i ___ drooping shoulders	**[9]** Glossopharyngeal
j ___ dysphagia	**[10]** Vagus
k ___ hoarseness	**[11]** Accessory
l ___ inability to look down	**[12]** Hypoglossal

These are terms you should know before proceeding to the post-test.   **MAJOR TERMS**

brainstem *(p. 348)*	pyramids *(p. 358)*
dura mater *(p. 348)*	pyramidal decussation *(p. 358)*
arachnoid *(p. 348)*	pons *(p. 358)*
pia mater *(p. 349)*	dorsal pons *(p. 358)*
ventricles *(p. 350)*	ventral pons *(p. 360)*
cerebral aqueduct *(p. 352)*	midbrain *(p. 361)*
cerebrospinal fluid (CSF) *(p. 352)*	cerebral peduncles *(p. 361)*
choroid plexuses *(p. 352)*	corpora quadrigemina *(p. 361)*
brainstem *(p. 356)*	nucleus ruber *(p. 361)*
nucleus *(p. 356)*	substantia nigra *(p. 361)*
ganglion *(p. 356)*	cerebellum *(p. 361)*
reticular formation *(p. 356*	cerebellar cortex *(p. 361)*
reticular activating system (RAS) *(p. 356)*	arbor vitae *(p. 361)*
medulla oblongata *(p. 358)*	cerebrum *(p. 363)*

ELECTROENCEPHALOGRAPHIC TECHNOLOGISTS AND TECHNICIANS

Electroencephalographic technologists operate electroencephalographs, the machines that record the electrical activity of the brain. The printed recording (trace) produced by the electroencephalograph is called an electroencephalogram (EEG). An EEG helps diagnose disease (such as brain tumors and strokes) and its effects on the brain. Technologists are experts with electroencephalographic equipment and know how to set it up for specific tests as well as how to use it with other equipment. Most electroencephalographic technologists work in hospitals, but some work with physicians who specialize in brain and nervous disorders.

Electroencephalographic technologists are often trained on the job (perhaps by first working as a technician), but formal training programs of one to two years are becoming more important for job opportunities. These formal programs are held at colleges, junior colleges hospitals, and vocational schools. The responsibilities of the technologist may also include laboratory management and the supervision of EEG technicians.

Electroencephalographic technicians work as assistants under the supervision of a technologist. They prepare patients for an EEG by helping them to relax and by applying the proper controls and electrodes for the particular recording. The arrangement of controls and electrodes varies depending on the disorder that is being tested, and is determined by the electroencephalographic technologist, electroencephalographic technician, and attending physician. Technicians are trained to recognize problems that could arise during an EEG, such as epileptic seizure.

Like technologists, technicians usually work in hospitals, but some work in private practices. Most technicians are trained on the job, although formal training programs are becoming more popular and desirable. These programs can be offered by colleges, junior colleges, hospitals, and vocational schools.

Based on terminology contained within this chapter, label the following right sagittal view of the brain:

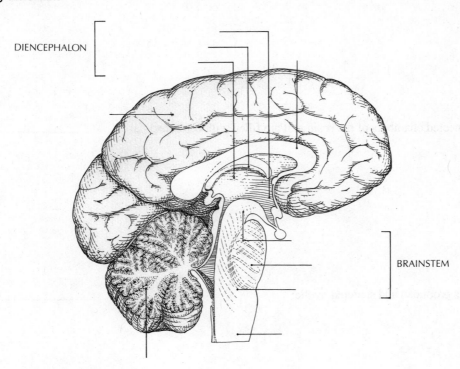

DIENCEPHALON

BRAINSTEM

After you have completed the following activities, compare your answers with those given at the end of this manual.

1 The blood-brain barrier:

   a is produced by the cells that make up the meninges.

   b is regulated by the microglia of the CNS.

   c is weaker in certain parts of the brain.

   d is uniform in its permeability throughout the CNS.

   e none of the above.

2 The reason the brain must have a constant blood supply is that:

   a energy cannot be produced without O_2.

   b CO_2 must be constantly removed to maintain proper pH.

   c little glucose is stored in nerve cells.

   d all of the above.

   e none of the above.

3 In the meninges, the large blood vessels pass through the:

   a pia mater.

   b dura mater.

   c subarachnoid space.

   d subdural space.

   e all of the above.

4 The hollow interconnected chambers of the brain that are filled with cerebrospinal fluid are the:

   a choroid plexuses.

   b pia mater.

   c foramina.

   d ventricles.

   e axons.

5 Cerebrospinal fluid is produced and secreted by the:

   a meninges.

   b ventricles.

   c choroid plexuses.

   d the brain.

   e none of the above.

6 Which part of the brain integrates all incoming nerve signals?

   a brainstem

   b cerebellum

   c thalamus

   d hypothalamus

   e pons

7 Ordinarily, CSF moves from the ventricles inside the brain to the subarachnoid space outside the brains.

   a true

   b false

8 The adult cerebrum is composed of _____ subdivisions.

   a two

   b four

   c six

   d eight

   e ten

9 Middle cerebellar peduncles are found within the: _____

   a brainstem.

   b medulla oblongata.

   c pons.

   d midbrain.

10 Deep sulci are known as: _____

   a gyri.

   b fissures.

11 In what order does CSF pass through the structures listed below? Put the letters in   _____

   the proper sequence.   _____

   a lateral ventricles   _____

   b subarachnoid space   _____

   c arachnoid villi   _____

   d fourth ventricle   _____

   e bloodstream   _____

   f interventricular foramen   _____

   g third ventricle

   h cerebral aqueduct

12 Ms. Harper is producing a large volume of urine (very dilute), has diminished   _____

   ADH secretion, and is unable to maintain constant body temperature. A physician

   suspects a brain tumor. In what part of the brain is the tumor probably located?

   _____.

13 Mr. Deaton exhibits tremor, lack of muscular coordination, and loss of muscular tone   _____

   on the right side of his body. What part of the brain is affected?_____

14 Indicate the main function of each of the following cranial nerves:

Name	Function
olfactory	_____
optic	_____
oculomotor	_____
trochlear	_____
trigeminal	_____
abducens	_____
acoustic	_____
glossopharyngeal	_____
vagus	_____
accessory	_____
hypoglossal	_____

15 Matching.

a ___ thalamus

      [1] Always lead to excitation of the
           effector organ.   _____

b ___ somatic fibers

      [2] Relay center for all sensations. _____

c ___ amphetamine

      [3] Stoppage of blood supply to the brain. _____

d ___ stroke

      [4] A sympathomimetic drug. _____

16 Matching.

a ___ afferent facial nerve

      [1] Hearing. _____

b ___ efferent facial nerve

      [2] Vagus. _____

c ___ taste

      [3] Accessory. _____

d ___ changes in the level of
      blood pressure.

      [4] Oculomtor. _____

e ___ shoulder movements

      [5] Taste. _____

f ___ smell

      [6] Hypoglossal. _____

g ___ cranial nerve III

      [7] Olfactory. _____

h ___ eye movements

      [8] Abducens. _____

i ___ tongue movements

      [9] Taste and other sensations of the
           tongue. _____

j ___ acoustic

      [10] Facial expression. _____

17 True or false.

i Cranial nerves are part of the CNS. _____

a true
b false

ii Clusters of neurons in the gray matter are called cell columns. _____

a true

b false

iii During embryological development, cells from the neural crest columns become _____
the afferent neurons.

a true
b false

iv The meninx that lies next to the skull is the pia mater. _____

a true

b false

v The meninges that surround the brain are continuous with those around the _____
spinal cord.

a true
b false

vi The superior saggital sinus drains blood from the brain.   _____

a true

b false

vii Microglia are present in all types of nerve tissue.   _____

a true

b false

viii Ependymal cells line the brain ventricles.   _____

a true

b false

ix The midbrain, pons, and medulla oblongata form the brainstem.   _____

a true

b false

x The cerebellum is chiefly involved with regulating acid-base balance.   _____

a true

b false

18 The colliculi are structures of the _____.   _____

19 The brain waves that occur in children and during emotional stress are   _____
    _____ waves.

20 The drug that is used to treat Parkinson's disease is _____ .   _____

21 A recurring pattern of seizures is characteristic of _____ .   _____

22 The primary somatic area of the brain is found in the _____ lobe. _____

23 Five homeostatic functions of the hypothalamus are _____ ,   _____
    _____ , _____ , _____ , _____ .   _____

                                                                                 _____

                                                                                 _____

                                                                                 _____

24 That part of the brain concerned with emotions and emotional expression is the   _____
    _____ .

25 The treelike white matter of the cerebellum is called the _____ . _____

The Autonomic Nervous System

15

Prefixes

ad-   toward
auto-   self
dys-   bad; disordered
intra-   within
moto-   move
para-   near; by the side of
pre-   before
post-   after
ram-   branch
soma-   body

Suffixes

-nomic   law
-vertebral   vertebra

Read Chapter 15 in the textbook, focusing on the major objectives. As you progress through it, with your textbook open, answer and/or complete the following. When you complete this exercise, you will have a thorough outline of the chapter.

DEVELOPING YOUR OUTLINE

I. Introduction (p. 394)

1 Where do three typical involuntary responses occur in the body?

    a

    b

    c

2 What are the two divisions of the ANS?

    a

    b

3 What is the primary function of the ANS?

OBJECTIVE 1

4 What is the basic function of the somatic motor (efferent) system?

OBJECTIVE 2

5 Describe the neuroanatomy of the ANS.

6 How are the ANS and somatic nervous system interrelated?

II. Structure of the Peripheral Autonomic Nervous System (p. 395)

OBJECTIVE 1

    1 What are the two divisions of the PNS?

      a

      b

A Anatomical Divisions of the Peripheral Autonomic Nervous System (p. 395)

    1 What are the two anatomical divisions (levels) of the peripheral ANS?

      a

      b

    2 Complete the following diagram on the divisions of the PNS:

B Preganglionic and Postganglionic Neurons (p. 395)

    1 The first neuron of the ANS is called **a** _____ and has its cell body located within the **b** _____ or **c** _____ and its axon within a **d** _____ or **e** _____ nerve. The cell body and dendrites of the postganglionic neuron lie within a **f** _____ outside the CNS. The unmyelinated axon of the postganglionic neuron terminates on **g** _____ , **h** _____ , or **i** _____ .

C Autonomic Ganglia (p. 396)

OBJECTIVE 4

    1 The three groups of autonomic ganglia are the:

       a

       b

       c

    2 The three largest prevertebral ganglia are the:

       a

       b

       c

D Plexuses (p. 396)

    1 Name the five major autonomic plexuses.

       a

       b

       c

       d

       e

E Sympathetic Division (p. 397)

OBJECTIVE 5

    1 When a preganglionic neuron of the sympathetic nervous system reaches a sympathetic trunk, it can take three different pathways. What are they?

       a

       b

       c

2 Identify the list in Column A with the correct division of the ANS in Column B. Answers may be used more than once.

Column A

a norepinephrine

b adrenergic fibers

c cholinergic fibers

d epinephrine

e acetylcholine

Column B

[1]   Released by sympathetic preganglionic neurons.

[2]   Another name for noradrenaline.

[3]   Released by sympathetic postganglionics.

[4]   Catechol-O-methyl transferase substrate.

[5]   MAO deactivates.

[6]   Parasympathetic division.

[7]   Released by somatic nerve fibers.

[8]   Sympathetic division.

[9]   Also secreted by the medulla of adrenal gland.

[10] A catecholamine.

[11] Reacts with alpha receptors.

[12] Reacts with beta receptors.

3 What are the major effects of sympathetic stimulation?

F Parasympathetic Division (p. 397)

  1 What are two other names for this division of the ANS?

    a

    b

  2 The preganglionic fibers from the cranial portion of the parasympathetic division are also known as the _____, whereas those of the sacral portion are the _____ .

  3 In the following table, indicate whether the sympathetic or parasympathetic fibers stimulate (↑) or inhibit (↓) the given body parts. Put a 0 (zero) if there is no innvervation.

Part of body	Sympathetic	Parasympathetic
iris of eye	a _____	b _____
sweat glands	c _____	d _____
heart rate	e _____	f _____
salivary glands	g _____	h _____

4 a The mode of action at the preganglionic synapse is called

_____ .

b The mode of action at the postganglionic synapse is called

_____ .

5 The effects of parasympathetic stimulation are (short-, long-) acting.

III. Central Autonomic Control Centers (p. 401)
   A Brainstem and Spinal Cord (p. 401)
      1 What role does the medulla oblongata play in the control of the ANS?

   B Hypothalamus (p. 401)
      1 Complete the table by indicating whether stimulation causes an increase
        (↑) or decrease (↓) in the given response.

Response	Anterior hypothalamus	Posterior hypothalamus
stroke volume	a _____	b _____
heart rate	c _____	d _____
intestinal peristalsis	e _____	f _____
blood pressure	g _____	h _____

      2 The anterior hypothalamus has a _____ role and sends
      impulses via the _____ division of the ANS, whereas the role
      of the posterior hypothalamus is _____ and uses the
      _____ division of the ANS.

   C Cerebral Cortex and Limbic System (p. 402)
      1 What are three structures of the limbic system?
      a

      b

      c

   D Visceral Reflex Arc (p.403)
      1 List, in order, the five components of an autonomic visceral reflex arc.
      a

      b

c

d

e

2 Give three examples of a spinal reflex arc.

a

b

c

3 What are three examples of a medullary reflex arc?

a

b

c

IV. Functions of the Autonomic Nervous System (p. 403)

    A Example of the Operation of the System: A Ski Race (p. 403)

        1 Explain how the sympathetic and parasympathetic systems function before, during, and after a downhill ski race.

    B Coordination of the Two Divisions (p. 404)

        1 What part of the brain maintains homeostasis in the sympathetic and parasympathetic systems?

    C Responses of Specific Organs (p. 404)

        1 Explain why the ANS does not control the basic activity of the organs it innervates.

V. When Things Go Wrong (p. 404)

    1 Briefly describe each of the following disorders involving the ANS:

        a Horner's syndrome

b autonomic dysreflexia

c achalasia

These are terms you should know before proceeding to the post-test.   **MAJOR TERMS**

autonomic nervous system (ANS) *(p. 394)*

visceral efferent motor system *(p. 394)*

autonomic ganglia *(p. 395)*

peripheral autonomic nervous system *(p. 395)*

thoracolumbar division *(p. 395)*

craniosacral division *(p. 395)*

preganglionic neuron *(p. 396)*

postganglionic neuron *(p. 396)*

paravertebral ganglia *(p. 396)*

sympathetic trunks *(p. 396)*

prevertebral ganglia *(p. 396)*

terminal ganglia *(p. 386)*

autonomic plexuses *(p. 396)*

sympathetic division *(p. 397)*

thoracolumbar outflow *(p. 397)*

white rami communicantes *(p. 397)*

gray rami communicantes *(p. 397)*

acetylcholine *(p. 397)*

norepinephrine *(p. 397)*

adrenergic system *(p. 397)*

parasympathetic division *(p. 397)*

cranial parasympathetic outflow *(p. 400)*

sacral parasympathetic outflow *(p. 400)*

cholinergic *(p. 400)*

visceral reflex arc *(p. 403)*

Horner's syndrome *(p. 404)*

autonomic dysreflexia *(p. 404)*

achalasia *(p. 404)*

SPEECH PATHOLOGIST AND ANESTHESIOLOGIST'S ASSISTANT (AA)

Speech pathologists work with people who have speech, language, and voice disorders that may result from a variety of sources, such as total or partial loss of hearing or mental retardation. Speech pathologists work directly with the patients. They test them to find the cause of the disorder, then develop a program of treatment for the patient. Some speech pathologists research the causes of communicative disorders, whereas others supervise clinical activities, teach, or are involved in administration. Many work in schools, whereas others work in clinics, government agencies, or private practice. A bachelor's or master's degree in speech and hearing programs is required and some states may also require teaching certificates to work in public schools. Requirements for licenses vary in different states. To advance in speech pathology, certification by the American Speech and Hearing Association is necessary.

The **anesthesiologist's assistant** (AA) functions under the direction of licensed and qualified anesthesiologists, principally in academic medical centers. The designated anesthesiologist is immediately available to the operating room to direct the anesthetic care in which an AA is involved. AAs assist the anesthesiologist in the collection of preoperative data, including taking an appropriate health history and performing the necessary physical examination and preoperative tasks, such as the insertion of intravenous and arterial lines, central venous pressure monitors, and special catheters. During surgery, AAs assist in airway management and drug administration for induction and maintenance of anesthesia; they administer supportive therapy such as intravenous fluids and vasodilators. AAs also perform supportive tasks in the recovery room, intensive care, or pain clinic care and can work in anesthesia monitoring services or in administrative functions and staff education. Further, AAs provide technical support. This includes first-level maintenance of anesthesia equipment, skilled operation of special monitors (i.e., echocardiographs, electroencephalographs, spectral analyzers, evoked-potential apparatuses, autotransfusion devices, mass spectrometers, intraaortic balloon pumps), operation and maintenance of bedside electronic computer-based monitors, supervised laboratory functions associated with anesthesia and operating room care, and cardiopulmonary resuscitation according to established protocols. In general, the duties of an anesthesiologist's assistant are distinct from those of the anesthesiologist and certified levels of understanding of physiology and chemistry, thereby giving them strong academic backgrounds for their work within the family of anesthesia providers.

The programs, which are master's degree level, are usually two years long. The programs require a baccalaureate degree, preferably in cell biology, chemistry, physics, mathematics, computer science, or physiology.

Based on the terminology within the chapter, label the following views of the spinal cord: **LABELING ACTIVITY**

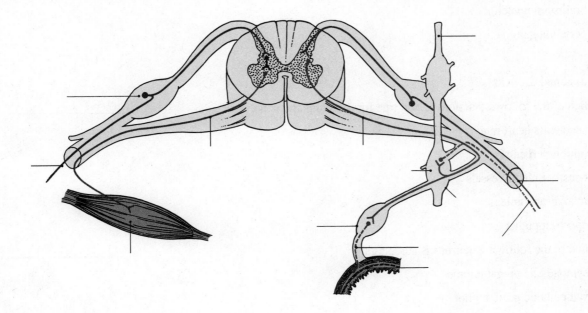

After you have completed the following activities, compare your answers with those given at the end of this manual.   **POST-TEST**

1 Complete the following table on the sites of acetylcholine action:

Sites	Autonomic nervous system division	Receptor
all autonomic ganglia	sympathetic and parasympathetic	a _____
all parasympathetic effector sites	b _____	muscarinic or nicotinic
c _____	sympathetic	muscarinic

2 Which of the following neurons liberates norepinephrine from its axonic terminals? _____

   a sympathetic postganglionics

   b sympathetic preganglionics

   c parasympathetic preganglionics

   d **a** and **b**

   e **a, b,** and **c**

3 The cell bodies of the preganglionic parasympathetic neurons are located in: _____

    a second, third, and fourth sacral segments.

    b brainstem nuclei.

    c vertebral ganglia.

    d **a** and **b**.

    e **a, b,** and **c.**

4 Which of the following actions is *not* true for the sympathetic nerves? _____

    a accelerate heart rate

    b inhibit peristalsis

    c constrict blood vessels

    d relax bronchioles

    e constrict pupils

5 Which of the following neurons is adrenergic? _____

    a sympathetic preganglionic

    b sympathetic postganglionic

    c parasympathetic preganglionic

    d parasympathetic postganglionic

    e all of the above

6 With respect to the autonomic nervous system, which action does *not* occur? _____

    a It functions in maintenance of homeostasis.

    b All preganglionic neurons are cholinergic.

    c The parasympathetic division is dominant during stressful situations.

    d Most autonomic fibers are tonically active.

    e The two divisions are antagonistic.

7 The parasympathetic division of the ANS stimulates the secretion of saliva. _____

    a true

    b false

8 The autonomic nervous system is a visceral effector system. _____

    a true

    b false

9 The control of visceral activities is a function of the cerebrum. _____

    a true

    b false

10 A mental process whereby an individual gets feedback signals about visceral body _____

functions is called _____ .

11 Most sympathetic postganglionic axons are _____ . _____

12 An autonomic pathway always consists of at least _____ neurons. _____

13 The ganglia located near the large abdominal arteries are called _____ _____

ganglia.

14 Only motor nerves comprise the autonomic nervous system. _____

    a true

    b false

15 Matching.

    a ___ autonomic dysreflexia   [1] Difficult reflexes.   _____

    b ___ Horner's syndrome   [2] Large intestine constricted.   _____

    c ___ achalasia   [3] Failure to relax.   _____

    d ___ Hirschsprung's disease   [4] Lesions in cervical ganglia T_1 to T_4.   _____

16 The main function of the autonomic nervous system is to promote _____. _____

17 Generally, the parasympathetic division is active when the body is at rest. _____

    a true

    b false

18 The _____ is the highest and main subcortical regulatory center of the _____

ANS.

19 Structures of the _____ system use the hypothalamus to express visceral _____

and behavioral activities related to self-preservation.

20 List, in order, the components of an autonomic visceral reflex arc. _____

    a _____

    b _____

    c _____

    d _____

    e

21 List the neural centers in the CNS that help control the ANS. _____

    a _____

    b _____

    c _____

    d _____

    e _____

    f

22 _____ receptors are either nicotinic or muscarinic. _____

23 Autonomic ganglia are classified as either _____ or _____ _____

ganglia in the sympathetic division and _____ ganglia in the parasym-

_____

pathetic division. Rows of paravertebral ganglia form the _____ trunks.   _____

24 The autonomic nervous system may also be called the _____ .   _____

25 Biofeedback control holds considerable promise as a means of treating some   _____

psychosomatic problems.

a true

b false

The Senses

16

Prefixes

aud-	to hear
baro-	pressure
chemo-	chemical
chromo-	color
cochlea-	snail
fore-	a pit
gusta-	taste
kin-	to move
lacr-	tears; weeping
lut-	yellow
mechano-	mechanical
oculi-	eye
olfact-	smell
opsi-	sight
oto-	ear
photo-	light
retin-	a network
scler-	hard
semi-	half
tect-	to cover
thermo-	heat
tympan-	drum
vitr-	glass

Suffixes

-ceptor	nerve apparatus
-itis	inflammation
-lith	a stone

DEVELOPING YOUR OUTLINE

Read Chapter 16 in the textbook, focusing on the major objectives. As you progress through it, with your textbook open, answer and/or complete the following. When you complete this exercise, you will have a thorough outline of the chapter.

I. Introduction (p. 408)

   1 Who first identified the five senses?

II. Sensory Reception (p. 408)

   1 Describe a receptor.

2 What are the two forms of functional activity that result from sensory input?

a

b

3 Describe a sensation.

4 Describe perception.

A Basic Characteristics of Sensory Receptors (p. 408)

OBJECTIVE 1

   1 What are three features that are basic to all sensory receptors?

     a

     b

     c

B Classification of Sensory Receptors (p. 408)

OBJECTIVE 2

   1 List and describe the four kinds of sensory receptors based on their location.

     a

     b

     c

     d

   2 List and describe five types of sensations associated with the general senses.

     a

     b

     c

d

e

3 What are five types of receptors based on the type of stimulus?

a

b

c

d

e

4 There are two kinds of nerve endings associated with sensory receptors.
What are they?

a

b

C Sensory Receptors and the Brain (p. 409)
   1 What is the main purpose of the senses?

   2 What is the role of the brain in sensory reception?

III. General Senses (p. 411)

OBJECTIVE 3

   A Light Touch (p. 411)
      1 List the three receptors involved with light touch.

      a

      b

      c

   B Touch-Pressure (Deep Pressure) (p. 411)
      1 What are the main receptors for deep pressure?

2 What is the two-point discrimination test used for?

C Vibration (p. 411)
    1 What types of receptors are involved in the sense of vibration?

D Heat and Cold (p. 411)
    1 Cutaneous receptors for heat and cold are _____ .

E Pain (p. 412)
    1 Describe three types of pain.

        a

        b

        c

    2 Which receptors are involved in pain perception?

    3 Which neural pathways are involved in the transmission of the pain impulse?

    4 Describe and give an example of referred pain.

    5 What are two explanations for referred pain?

        a

        b

    6 What is phantom pain?

OBJECTIVE 4

F Proprioception (p. 413)
    1 Where are proprioception receptors located?

OBJECTIVE 5

G Itch and Tickle (p. 413)
    1 What is the difference between an itch and a tickle?

    2 Where are these receptors located?

H Sterognosis (p. 413)

    1 What is stereognosis?

I Corpuscles of Ruffini and Bulbous Corpuscles (of Krause) (p. 413)

    1 Where are corpuscles of Ruffini located?

    2 List three functions of corpuscles of Ruffini.

        a

        b

        c

    3 Where are bulbous corpuscles (of Krause) located?

    4 What type of receptors are bulbous corpuscles?

J Neural Pathways for General Senses (p. 413)

    1 Complete the following table on neural pathways for general sense:

Function	Neural pathways
touch	**a** _____
	b _____
	c _____
d _____	dorsal column-medial lemniscal pathway, spinocervicothalamic pathway
temperature	**e** _____
pain	**f** _____
	g _____
proprioception	**h** _____
	i _____
sensations from face	**j** _____

IV. Taste (Gustation) (p. 414)

OBJECTIVE 6

  A Structure of Taste Receptors (p. 414)

    1 List the three main types of papillae.

      a

      b

      c

  B Basic Taste Sensations (p. 415)

    1 What are the four presumed basic taste sensations?

      a

      b

      c

      d

  C Neural Pathways for Gustation (p. 416)

    1 Describe the neural pathway for taste.

V. Smell (Olfaction) (p. 416)

OBJECTIVE 7

  A Structure of Olfactory Receptors (p. 416)

    1 What are the three types of cells found within the olfactory epithelium?

      a

      b

      c

  B Neural Pathways for Olfaction (p. 417)

    1 Describe the neural pathway for olfaction.

VI. Hearing and Equilibrium (p. 418)

OBJECTIVE 8

  A Structure of the Ear (p. 418)

    1 List the major parts of the external ear.

    2 What is the function of the cerumen?

3 List the major parts of the middle ear.

4 What is the major function of the Eustachian tube?

5 List the major parts of the inner ear.

6 List the three spiral ducts of the cochlea.

   a

   b

   c

B Neural Pathways for Hearing (p. 423)
   1 Describe the neural pathways for hearing.

C Vestibular Apparatus and Equilibrium (p. 423)

> OBJECTIVE 9

   1 Compare and contrast dynamic and static equilibrium.

   2 What are the two types of sensory hairs present in hair bundles?

      a

      b

   3 What are the receptor organs for static equilibrium called?

   4 Changes in the direction of head movements are interpreted by what structures?

   5 Describe the mechanisms involved in static and dynamic equilibrium.

D Neural Pathways for Equilibrium (p. 426)

> OBJECTIVE 10

   1 Describe the neural pathways for equilibrium.

VII. Developmental Anatomy of the Ear (p. 426)
   1 Describe how the ear develops.

VIII. **Vision** (p. 426)

    A Structure of the Eyeball (p. 426)

        1 What are the three layers of the eyeball?

            a

            b

            c

        2 Give a function of each of the following:

            a sclera

            b cornea

            c ciliary body

            d lens

            e iris

            f pupil

            g neuroretina

            h macula lutea

            i fovea

            j rods

            k cones

        3 What are the three types of color cones?

            a

b

c

4 List the three cavities of the eyeball.

a

b

c

B Accessory Structures of the Eye (p. 433)

1 List the major function of the following:

a bony orbit

b eyelids

c eyelashes

d eyebrows

e conjunctiva

2 What three structures make up the lacrimal apparatus?

a

b

c

3 List four functions of tears.

a

b

c

d

4 Describe the action of the six sets of eye muscles.

C Physiology of Vision (p. 435)
1 What are the five phases of the visual process?

a

b

c

d

e

D Neural Pathways for Vision (p. 435)

OBJECTIVE 12

1 The field of vision for each eye is divided into what two parts?

a

b

2 What is the significance of the optic chiasma?

3 How does the eye adapt to light and darkness?

4 What is the second visual system?

IX. Developmental Anatomy of the Eye (p. 437)
1 Describe how the eye develops.

X. When Things Go Wrong (p. 438)

OBJECTIVE 13

1 Briefly discuss the following disorders:

a otosclerosis

b labyrinthitis

c Ménière's disease

d motion sickness

e otitis media

f detached retina

g cataract

h conjunctivitis

i glaucoma

These are terms you should know before proceeding to the post-test   **MAJOR TERMS**

receptors *(p. 408)*

exteroceptors *(p. 409)*

teleceptors *(p. 409)*

interoceptors *(p. 409)*

proprioceptors *(p. 409)*

general senses *(p. 409)*

thermoreceptors *(p. 409)*

nociceptors *(p. 409)*

chemoreceptors *(p. 409)*

photoreceptors *(p. 409)*

mechanoreceptors *(p. 409)*

baroreceptors *(p. 409)*

free nerve endings *(p. 409)*

encapsulated endings *(p. 409)*

tactile (Merkel's) corpuscles *(p. 411)*

tactile (Meissner's) corpuscles *(p. 411)*

lamellated (Pacinian) corpuscles *(p. 411)*

vibration *(p. 411)*

specialized free nerve endings *(p. 412)*

referred pain *(p. 412)*

phantom pain *(p. 412)*

proprioception *(p. 413)*

itch *(p. 413)*

tickle *(p. 413)*

stereognosis *(p. 413)*

corpuscles of Ruffini *(p. 413)*

bulbous corpuscles (of Krause) *(p. 413)*

gustation *(p. 414)*

papillae *(p. 414)*

taste buds *(p. 414)*

olfaction *(p. 416)*

olfactory receptor cells *(p. 416)*

optic vesicles *(p. 437)*

optic stalk *(p. 437)*

lens placode *(p. 438)*

optic cup (p. 438)

otosclerosis *(p. 438))*

labyrinthitis *(p. 438)*

Ménière's disease *(p. 438)*

motion sickness *(p. 438)*

otitis media *(p. 439)*

detached retina *(p. 439)*

cataract *(p. 439)*

conjunctivitis *(p. 439)*

glaucoma *(p. 439)*

Based on the terminology contained within this chapter, label the following figures: **LABELING ACTIVITY**

1   Section through the ear.

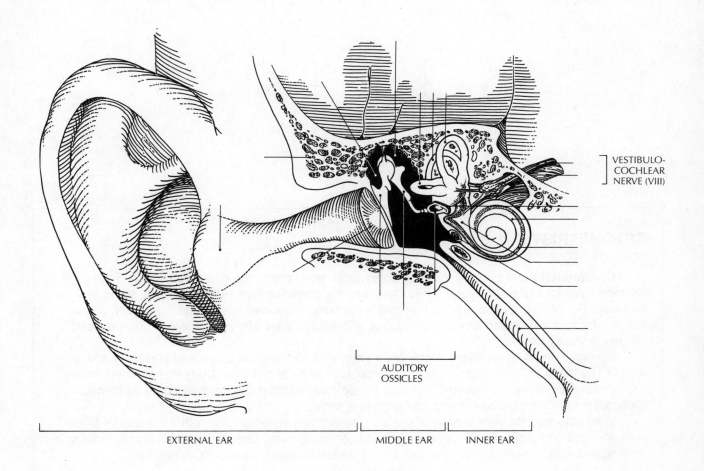

VESTIBULO-
COCHLEAR
NERVE (VIII)

AUDITORY
OSSICLES

EXTERNAL EAR   MIDDLE EAR   INNER EAR

2   Section through the eyeball.

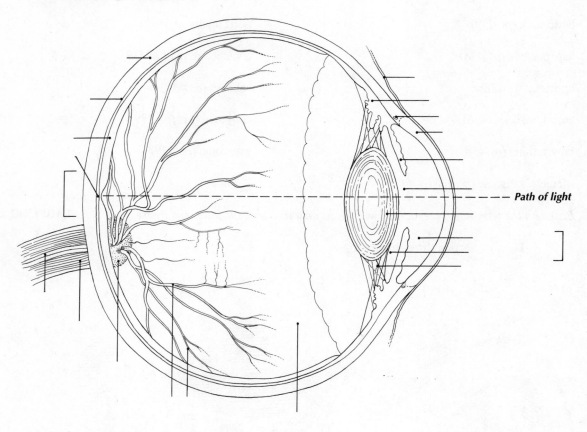

Path of light

OPTOMETRIST

Optometrists test eyes for focusing ability, color perception, and depth perception. They also examine eyes for signs of disease. They can prescribe corrective lenses and can treat some eye diseases. For serious eye conditions that may require more involved treatments or surgery, optometrists refer their patients to an ophthalmologist, a medical doctor who specializes in diseases and surgery of the eye.

Optometrists must receive a doctor of optometry (D. O.) from an accredited university and are licensed by the state. The degree programs entail four years of graduate study that combine theory and clinical experience. To become licensed, the optometrist must pass a state board examination. Some states also require continuing education in optometry.

Most optometrists work in general practice, sometimes in conjunction with an optician's office. (An optician fills prescriptions for eyeglasses and contact lenses.) Optometrists can also specialize in working with children or the elderly or can work in industrial safety, research, or teaching.

After you have completed the following activities, compare your answers with those given at the end of this manual.

1 Generation of an action potential in a sensory neuron requires that:
   a the receptor cell membrane be sensitive to the stimulus.
   b a receptor potential be generated.
   c a great deal of energy be present in the stimulus.
   d a and b
   e a, b, and c

        _____

2 Pressure receptors in the skin are called:
   a tactile corpuscles (of Meissner).
   b free nerve endings.
   c lamellated (Pacinian) corpuscles.
   d corpuscles of Ruffini.
   e bulbous corpuscles (of Krause).

        _____

3 Which of the following is an example of a mechanoreceptor?
   a baroreceptors
   b stretch receptors in the stomach
   c stretch receptors in the intestine
   d a and b.
   e a, b, and c.

        _____

4 Which of the following is *not* part of the inner ear?
   a osseous labyrinth
   b vestibule
   c semicircular canal
   d tympanic membrane
   e cochlea

        _____

5 Which of the following is involved in the sense of dynamic equilibrium?
   a utricle
   b saccule
   c semicircular canal
   d a and b
   e a, b, and c

        _____

6 Accommodation is possible as a result of the lens's:

   a curvature.

   b position.

   c elasticity.

   d pigments.

   e opacity.

7 Rods have a higher degree of _____ than do cones.

   a color sensitivity

   b light sensitivity

   c resolution

   d magnification

   e none of the above

8 A _____ is any device that converts one form of energy to another.

9 Although most neurons can probably respond to mechanical distortion by generating a membrane potential, the receptor membranes of _____ cells are highly specialized for such reception.

10 The most thoroughly studied of all proprioceptors are the _____ spindles.

11 One principal structure of the mammalian middle ear is the auditory _____.

12 The sense of hearing is localized within the _____ .

13 The actual organ of hearing is the organ of _____ .

14 In sound detection, vibrations in the fluid of the cochlea of the ear are detected by sensory receptors called _____ cells.

15 Matching.

   a ___ exteroceptor   [1] Movements.

   b ___ interoceptor   [2] Odors.

   c ___ proprioceptor   [3] pH.

16 Matching.

   a ___ phasic receptor   [1] Molecular detection of chemical stimuli.

   b ___ tonic receptor   [2] A short burst of action potentials.

   c ___ chemoreception   [3] Continual, steady, intense stimulation.

17 Matching.

  a ___ tactile corpuscle   [1] Temperature.   _____
         (of Meissner)

  b ___ corpuscles of Ruffini   [2] Fine touch.   _____

  c ___ bulbous corpuscles   [3] Pain.   _____
         (of Krause)

18 Matching.

  a ___ statoliths   [1] A funnel.   _____

  b ___ auricle   [2] Loose crystals of calcium salt.   _____

  c ___ tympanum   [3] A membrane.   _____

19 Each sensory axon _____ . Which is false?   _____

  a carries information for only one sensory modality.

  b reports information from only one small location on the body.

  c conducts information concerning the intensity of stimulation.

  d makes synaptic contact with a central neuron.

  e typically responds to more than one type of sensory stimulation.

20 Information is gathered by a sensory nerve ending in a manner analogous to   _____

   _____ and then transmitted along the axon in a manner analogous to   _____

   _____ .

  a AM (amplitude modulation), FM (frequency modulation)

  b AC (alternating current), DC (direct current)

  c digital recording, analog recording

  d parallel processing, serial processing

  e RAM (ready access memory), ROM (read only memory)

21 The so-called five senses—sight, hearing, smell, taste, and touch—entirely ignore   _____
   the importance of:

  a exteroceptors.

  b interoceptors.

  c chemoreceptors.

  d special receptors.

  e transducers.

22 Proprioception is absolutely essential to effective locomotion. Proprioceptors are   _____

   stimulated by _____ . Which is false?

   a stretching any skeletal muscle.

   b tension applied to tendons.

   c tension applied to ligaments.

   d relation of a skeletal muscle to another muscle.

   e the weight of a limb lengthening its skeletal muscles.

23 Comparing the vertebrate eye with a 35-mm reflex camera, the _____   _____

   performs the same function as the _____ .   _____

   a choroid layer, film

   b iris, diaphragm

   c ciliary body, camera back

   d accommodation mechanism, shutter

   e cornea, lens cap

24 Although most body tissues are pigmented and opaque, in the vertebrate eye, the   _____

   living cells of the _____ are transparent to light.

   a cornea, lens, and retinal neurons

   b iris, cornea, and vitreous humor

   c ciliary body, aqueous humor, and vitreous humor

   d lens, retina, and choroid layer

   e area centralis, fovea, and retina

25 An inflammation of the middle ear is called _____ .   _____

The Endocrine System

17

Prefixes

adeno-	gland
anti-	against
diure-	urinate
endo-	within
exo-	outside
folli-	a bag
glyco-	sweet
gono-	a seed
hormon-	to set in motion
lacto-	milk
modul-	measure
neuro-	nerve
para-	near
poly-	many
thalam-	chamber
trop-	turn; change
troph-	to nourish

Suffixes

-algesia	pain
-crin	to secrete
-gon	generate; seed
-physis	growth
-tropic	influencing

Read Chapter 17 in the textbook, focusing on the major objectives. As you progress through it, with your textbook open, answer and/or complete the following. When you complete this exercise, you will have a thorough outline of the chapter.

DEVELOPING YOUR OUTLINE

I. Introduction (p. 446)

1 Describe the major functions of the endocrine system.

2 Describe the locations of the major endocrine glands.

> OBJECTIVE 1

3 Define a hormone.

> OBJECTIVE 2

II. Pituitary Gland (Hypophysis) (p. 446)

> OBJECTIVE 3

1 Why is the pituitary also called the hypophysis?

2 Where is this gland located?

3 What structure protects this gland?

4 What are its two lobes called?

   a

   b

5 What connects it to the hypothalamus?

A Relationship Between the Pituitary and Hypothalamus (p. 448)

   1 What are two types of secretions produced by the hypothalamus?

      a

      b

   2 What are the connecting links between the pituitary and hypothalamus? What is the function of the infundibular stalk? The hypothalamic-hypophyseal portal system?

   3 Does the neurohypophysis manufacture any hormones? Explain.

III. Thyroid Gland (p. 448)

OBJECTIVE 4

   1 In the space below, draw the thyroid gland and label completely.

   2 What is unique about the thyroid gland?

IV. Parathyroid Glands (p. 449)

   1 How many parathyroids does one have?

2 List the two cell types present in the parathyroid glands and give the secretion of each.

a

b

V. Adrenal Glands (p. 449)

1 The two anatomical parts of the adrenal gland are the _____

and _____ .

A Adrenal Cortex (p. 450)

OBJECTIVE 5

1 What are the three classes of general steroids produced by the adrenal cortex

a

b

c

2 From the outside to the inside, what are the three distinct zones of the cortex?

a

b

c

B Adrenal Medulla (p. 450)

OBJECTIVE 6

1 Name the two hormones secreted by the adrenal medulla.

a

b

2 What is the function of chromaffin cells?

VI. Pancreas (p. 451)

OBJECTIVE 7

1 What are the three anatomical parts of the pancreas?

a

b

c

2 Why is the pancreas called a mixed gland?

3 What do each of the following cell types produce?

a Alpha cells

b Beta cells

c Delta cells

d F cells

VII. Gonads (p. 451)

1 What are the male sex hormones called? Female sex hormones?

VIII. Other Sources of Hormones (p. 451)

A Kidneys (p. 451)

1 What is the function of the renin?

B Pineal Gland (p. 451)

OBJECTIVE 8

1 What are two other names for this gland?

a

b

2 What hormone has been isolated from the pineal gland? What is its possible function?

C Thymus Gland (p. 453)

OBJECTIVE 9

1 When is the thymus gland the most active?

2 What is the general function of the thymus?

D Heart (p. 453)
  1 What hormone is produced by the heart?

E Digestive System (p. 453)
  1 What are the functions of:

  a gastrin

  b secretin

  c cholecystokinin

F Placenta (p. 453)
  1 What hormones are produced by the placenta?

IX. The Effects of Aging on the Endocrine System (p. 453)
  1 What are several changes that occur with age and the endocrine system?

X. Developmental Anatomy of the Pituitary Gland (p. 456)
  1 Describe how the pituitary gland develops.

XI. When Things Go Wrong (p. 456)

  1 Match the disorder with the hormonal imbalance.

  a ___ giantism   [1] Oversecretion of growth hormone
                                           during adulthood.

  b ___ acromegaly   [2] Overactive thyroid.

  c ___ pituitary dwarf   [3] Hyperparathyroidism.

  d ___ polyuria   [4] Overproduction of glucocorticoids.
       (diabetes insipidus)

  e ___ Graves' disease   [5] Oversecretion of growth hormone
                                           during skeletal development.

  f ___ goiter   [6] Undersecretion of growth
                                           hormone.

  g ___ cretinism   [7] Undersecretion of ADH.

  h ___ osteoporosis   [8] Underactive thyroid in adult.

  i ___ Cushing's disease   [9] Underactive thyroid in child.

  j ___ Addison's disease   [10] Underactivity of the adrenal cortex.

  k ___ diabetes mellitus   [11] Not enough insulin.

  l ___ insulin, exercise, diet   [12] Used to treat diabetes mellitus.

These are terms you should know before proceeding to the post-test.　　　　　**MAJOR TERMS**

endocrine system *(p. 446)*

endocrine glands *(p. 446)*

hormone *(p. 446)*

target cells/pituitary gland (hypophysis) *(p. 446)*

adenohypophysis *(p. 446)*

neurohypophysis *(p. 446)*

infundibulum *(p. 446)*

hypothalamus *(p. 448)*

hypothalamic-hypophyseal portal system *(p. 448)*

thyroid gland *(p. 448)*

parathyroid glands *(p. 449)*

adrenal glands *(p. 449)*

adrenal cortex *(p. 450)*

adrenal medulla *(p. 450)*

pancreas *(p. 451)*

pancreatic islets *(p. 451)*

pineal gland *(p. 451)*

melatonin *(p. 451)*

thymus gland *(p. 453)*

atriopeptin *(p. 453)*

secretin *(p. 453)*

gastrin *(p. 453)*

cholecystokinin *(p. 453)*

placenta *(p. 453)*

giantism *(p. 457)*

acromegaly *(p. 457)*

pituitary dwarf *(p. 457)*

diabetes insipidus *(p. 458)*

hyperthyroidism *(p. 458)*

hypothyroidism *(p. 458)*

cretinism *(p. 458)*

myxedema *(p. 458)*

hypoparathyroidism *(p. 458)*

hyperparathyroidism *(p. 458)*

Cushing's disease *(p. 458)*

adrenogenital syndrome *(p. 459)*

Addison's disease *(p. 459)*

diabetes mellitus *(p. 459)*

hypoglycemia *(p. 459)*

Based on the terminology contained within the chapter, label the following endocrine glands:

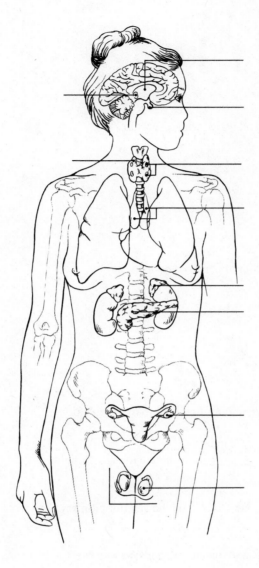

After you have completed the following activities, compare your answers with those given at the end of this manual.

1 Which of the following is a physiological effect of hormones?

   a control body growth

   b regulation of reproduction

   c maintenance of homeostasis

   d integration and coordination

   e all of the above

_____

2 Which of the following is *not* a gonadotropin?

   a luteinizing hormone

   b follicle-stimulating hormone

   c lactogenic hormone

   d adrenocorticotropin

   e **c** and **d**

3 The secretory activity of the thyroid gland varies with respect to:

   a the season of the year.

   b pregnancy.

   c stress.

   d trauma.

   e all of the above.

4 Of the following, which is *not* a symptom of corticoid deficiency?

   a muscular weakness

   b increased body temperature

   c hypoglycemia

   d hypotension

   e dehydration

5 _____ glands release their secretions directly into the circulatory system.

6 Until released into the circulation, _____ is bound to thyroglobulin, a protein.

7 _____ , a glycoprotein hormone secreted by the anterior pituitary, controls thyroid activity.

8 The adrenal glands in humans consist of two distinct portions: an outer _____ and an inner medulla.

9 Epinephrine is also called _____ .

10 Matching.

   a ___ ADH   [1] Somatotropin.

   b ___ growth hormone   [2] Vasopressin.

   c ___ affects mammary glands   [3] Prolactin.

11 Matching.

   a ___ noradrenaline   [1] A mineralocorticoid.

   b ___ cortisone   [2] A catecholamine.

   c ___ aldosterone   [3] Glucocorticoid.

12 Matching.

    a ___ gastrin   [1] Stimulates gonadal tissue.   _____

    b ___ secretin   [2] Produced by cell of the stomach wall.   _____

    c ___ FSH   [3] Produced by cells lining the   _____

                                       duodenum.

13 Matching.

    a ___ estrogen   [1] Causes growth of the endometrium.   _____

    b ___ vasopressin   [2] Acts on the smooth muscles of blood   _____

                                         vessels.

    c ___ oxytocin   [3] Causes uterine contraction.   _____

The Cardiovascular System: The Blood

18

Prefixes

erythro-	red
hema-	blood
hemo-	blood
leuko-	white
mega-	large
phil-	loving
poly-	many
poie-	make; produce
thrombo-	clot

Suffixes

-chrom	color
-cyte	cell
-emia	blood
-gen	creates; forms
-morph	shape
-phil	loving
-poiesis	make; produce
-stasis	standing still

Read Chapter 18 in the textbook, focusing on the major objectives. As you progress through it, with your textbook open, answer and/or complete the following. When you complete this exercise, you will have a thorough outline of the chapter.

DEVELOPING YOUR OUTLINE

I. Introduction (p. 465)

1 How is blood classified?

2 List the four parts of the hematologic system.

a

b

c

d

II. Functions of Blood (p. 465)

1 List three important functions of blood.

a

b

c

OBJECTIVE 1

III. Properties of Blood (p. 465)

1 Describe these quantitative characteristics of blood.

   a percent of body weight

   b liters in male

   c liters in female

   d specific gravity

   e pH

   f temperature

   g viscosity

IV. Components of Blood (p. 465)

  A Plasma (p. 465)

    1 What is the major component of plasma?

    2 The total protein component of blood plasma can be divided into the

      _____ , _____ , and _____ .

    3 What are the three classes of globulins?

      a

      b

      c

    4 A plasma protein essential for blood clotting is _____ .

    5 When all proteins are removed from plasma, the remaining liquid is

      called _____ .

    6 The major cation of plasma is the _____ , whereas the major

      anion is the _____ .

    7 The only source of energy for red blood cells is _____ .

8 What are the four forms of lipids that are found in the plasma?

a

b

c

d

9 The three principal gases dissolved in plasma are:

a

b

c

B Red Blood Cells (Erythrocytes) (p. 467)

OBJECTIVE 4

1 Draw a mature erythrocyte.

2 What advantage does the above shape offer this cell?

3 What accounts for the red color of erythrocytes?

4 Specifically, where does an erythrocyte carry oxygen?

OBJECTIVE 5

5 Oxygenated hemoglobin is called _____ .

6 How is carbon dioxide transported in the blood?

a

b

c

7 Where are fetal blood cells produced in a 4-month fetus?

_____ In a 7-month fetus? _____ In a newborn

infant? _____

8 List the stages and cell types involved in erythropoiesis.

OBJECTIVE 6

9 Why is the life span of an erythrocyte only about 120 days?

OBJECTIVE 7

10 Where and how are aged erythrocytes destroyed?

C White Blood Cells (Leukocytes) (p. 468)

OBJECTIVE 8

   1 What is a normal leukocyte count?

   2 What is the production of white blood cells called?

   3 State a general function of leukocytes.

   4 What are two categories of leukocytes?

   5 Contrast the five types of leukocytes by completing the following table:

OBJECTIVE 9

Cell type	Approximate number	Origin	Description	Function	Classification
neutrophil	_____	_____	_____	_____	_____
eosinophil	_____	_____	_____	_____	_____
basophil	_____	_____	_____	_____	_____
monocyte	_____	_____	_____	_____	_____
lymphocyte	_____	_____	_____	_____	_____

   6 Complete the following stages of granulocyte development:

bone marrow → a _____ → promyelocyte →

b _____ , _____ , _____ , →

metamyelocytes → c _____ → basophils, neutrophils,

d _____ .

7 How do lymphocytes differ from other leukocytes?

8 Contrast the two types of lymphocytes.

   a

   b

9 What are plasma cells?

D Platelets (Thrombocytes) (p. 472)　　　　　　　　　　　　　　　　　　　　OBJECTIVE 10
   1 Complete the following information about platelets:
   a main function

   b number present/mm^3 of blood

   c size and shape

   d origin

   e two secretory products

V. Hemostasis: The Prevention of Blood Loss (p. 473)　　　　OBJECTIVE 11
   A Vasoconstrictive Phase (p. 473)
   1 What causes vasoconstriction?

   B Platelet Phase (p. 473)
   1 What does platelet aggregation form?

   2 Why is platelet aggregation important?

   C Hemostasis and the Nervous System (p. 473)
   1 How does the nervous system contribute to hemostasis?

VI. When Things Go Wrong (p. 473)
   A Anemias (p. 473)
   1 List the causes of the following anemias:
   a hemorrhagic

   b iron-deficiency

c aplastic

d hemolytic

e thalassemia

f pernicious

g sickle-cell

B Hemophilia (p. 474)
   1 What is hemophilia?

C Leukemia (p. 474)
   1 What are the two major forms of leukemia?
      a

      b

   2 How do the various leukemias differ?

These are the terms you should know before proceeding to the post-test. **MAJOR TERMS**

blood *(p. 465*   erythropoiesis *(p. 468)*

plasma *(p. 465)*   stem cells (hemocytoblasts) *(p. 468)*

formed elements *(p. 465)*   reticulocyte *(p. 468)*

plasma proteins *(p. 466)*   leukocytes *(p. 468)*

albumins *(p. 466)*   leukopoiesis *(p. 468)*

globulins *(p. 466)*   granulocytes *(p. 468)*

fibrinogen *(p. 466)*   neutrophils *(p. 469)*

serum *(p. 466)*   eosinophils *(p. 469)*

electrolytes *(p. 466)*   basophils *(p. 469)*

erythrocytes *(p. 467)*   agranulocytes *(p. 469)*

hemoglobin *(p. 467)*   monocytes *(p. 469)*

lymphocytes *(p. 469)*

B cells *(p. 469)*

T cells *(p. 469)*

platelets (thrombocytes) *(p. 472)*

megakaryoblasts *(p. 472)*

coagulation *(p. 472)*

hemostasis *(p. 473)*

platelet aggregation *(p. 473)*

anemia *(p. 473)*

hemorrhagic anemia *(p. 473)*

iron-deficiency anemia *(p. 473)*

aplastic anemia *(p. 473)*

hemolytic anemia *(p. 474)*

pernicious anemia *(p. 474)*

sickle-cell anemia *(p. 474)*

hemophilia *(p. 474)*

leukemia *(p. 474)*

bone marrow transplantation *(p. 475)*

SPECIALIST IN BLOOD BANK TECHNOLOGY

Specialists in blood bank technology perform both routine and specialized tests in blood bank immunohematology. They use methodology that conforms with the Standards for Blood Banks and Transfusion Services of the American Association of Blood Banks. These specialists demonstrate a superior level of technical proficiency and problem-solving ability in such areas as (1) testing for blood group antigens, compatibility, and antibody identification; (2) investigating abnormalities such as hemolytic disease of the newborn, hemolytic anemias, and adverse responses to transfusion; (3) supporting physicians in transfusion therapy, including patients with coagulopathies or candidates for homologous organ transplant; (4) blood collection and processing, including selecting donors, drawing and typing blood, and performing pretransfusion tests to ensure the safety of the patient. Supervision, management, and/or teaching comprise a considerable part of the specialist's responsibilities.

Specialists in blood banking work in many types of facilities, including community blood centers, private hospital blood banks, university-affiliated blood banks, transfusion services, and independent laboratories. They may also be a part of a university faculty. The work schedules of these specialists vary; they may be required to work weekends, nights, and on emergency calls, depending on the facility. Qualified specialists may advance to supervisory or administrative positions, or move into teaching or research activities. The criteria for advancement in this field are experience, technical expertise, and completion of advanced education courses.

After you have completed the following activities, compare your answers with those given at the end of this manual.

POST-TEST

1 The liver plays several important roles in blood clotting, including production of: _____

   a fibrinogen.

   b platelets.

   c prothrombin.

   d bile salts, which are required for the absorption of vitamin K.

   e all but **b.**

2 Which of the following is true of plasma?

   a specific gravity of 1.03

   b pH of 7.4–7.5

   c contains inorganic constituents

   d makes of 55 percent of blood volume

   e all of the above

3 Neutrophils are:

   a precursors of platelets.

   b agranular.

   c leukocytes.

   d all of the above.

   e none of the above.

4 Which of the following is *not* true of RBC?

   a They are also called erythrocytes.

   b They are shaped like biconcave disks.

   c They contain hemoglobin.

   d Their density is 5 to 6 cells/mm^3.

   e Both **a** and **d**.

5 The blood volume in an adult averages approximately:

   a 1 liter.

   b 3 liters.

   c 5 liters.

   d 7 liters.

   e 10 liters.

6 Which of the following is *not* true of erythrocytes?

   a They are destroyed by the spleen.

   b Their life span is approximately 120 days.

   c They are manufactured in long bones of the body.

   d One of their functions is phagocytic activity.

7 Which of the following vitamins is necessary for the manufacture of hemoglobin?

   a A

   b K

   c B_{12}

   d B_1

   e B_6

8 The inflammatory process is characterized by all of the following *except:*

   a tissue swelling.

   b redness.

   c tissue tenderness.

   d erythrocytes phagocytose invaders.

   e both **a** and **b.**

9 Which of the following is the most numerous WBC?

   a eosinophil

   b neutrophil

   c monocyte

   d lymphocyte

   e T cell

10 In the adult, the synthesis of heme occurs in the:

   a bone marrow.

   b liver.

   c spleen.

   d lymph glands.

   e none of the above.

11 Which of the following is a vascular factor in hemostasis?

   a increase in venous pressure.

   b vasoconstriction in order to reduce blood flow.

   c decrease in blood pressure

   d vasodilation to bring more clotting factors to an area

   e a and b

12 Which of the following are the most abundant plasma proteins?

   a albumins

   b globulins

   c fibrinogens

13 Which of the following is needed to form a mature erythrocyte?

   a vitamin B_{12}

   b folic acid

   c amino acids

   d all of the above

   e none of the above

14 Which of the following is a cause of anemia?

    a deficiency of iron

    b bone marrow failure

    c excessive blood loss

    d low erythropoietin secretion

    e **a, b,** and **c**

15 Blood cells originate from a single cell type known as a(n):

    a ABC

    b stem cell

    c base cell

    d CFU

    e none of the above

16 Which of the following is the first to produce blood cells in the fetus?

    a liver

    b bone tissue

    c heart

    d yolk sac

    e kidneys

17 Which of the following is *not* a myeloid cell?

    a T lymphocyte

    b B lymphocyte

    c erythrocyte

    d monocyte

    e granulocyte

The Cardiovascular System: The Heart

19

Prefixes

baro-	pressure
bi-	two
brady-	slow
cardi(o)-	heart
diastol-	relax; stand apart
ectop-	displaced
epi-	above
iso-	equal
metr-	measure
myo-	muscle
peri-	around
pulmo-	lung
sclero-	hard
systol-	contract; stand together
tachy-	fast
tri-	three

Suffixes

-cusp	point
-lunar	moonlike
-stasis	standing

Read Chatper 19 in the textbook, focusing on the major objectives. As you progress through it, with your textbook open, answer and/or complete the following. When you complete this exercise, you will have a thorough outline of the chapter.

DEVELOPING YOUR OUTLINE

I. Introduction (p. 479)

1 What is meant by the statement that blood flows within a closed system of vessels?

2 What is the main function of the heart?

3 Why is the heart considered to be a double pump?

4 What is the difference between pulmonary and systemic circulation?

II. Structure of the Heart (p. 479)

OBJECTIVE 1

A Location of the Heart (p. 479)

    1 List the four surfaces of the heart.

      a

      b

      c

      d

    2 Describe the apex and base of the heart.

    3 Sketch the position and location of the heart.

B Covering of the Heart: Pericardium (p. 479)

OBJECTIVE 2

    1 The protective covering of the heart is called the _____ .

    2 Describe the placement of the following layers:

      a epicardium

      b parietal pericardium

      c visceral pericardium

    3 What is the function of pericardial fluid?

C Wall of the Heart (p. 481)

    1 List the layers of the heart from the outside to the inside.

D Cardiac Skeleton (p. 482)

    1 What is the function of the cardiac skeleton?

E Chambers of the Heart (p. 483)

OBJECTIVE 3

    1 What does the septum do?

2 Describe the structure and function of the following:

a atrium

b fossa ovalis

c musculi pectinati

d trabeculae carneae

e chordae tendineae

f coronary sulcus

g anterior interventricular sulcus

h right heart

i left heart

F Valves of the Heart (p. 483)

1 Test your understanding of heart valves by completing this matching exercise. More than one answer may be possible.

a ___ AV valves

[1] Right semilunar valve.

b ___ SV valves

[2] Permit the flow of blood from the atria to the ventricles.

c ___ cusps

[3] Permit the flow of blood to the pulmonary artery and aorta.

d ___ bicuspid valve

[4] The left AV valve.

e ___ mitral valve

[5] Flaps.

f ___ chordae tendineae

[6] Prevents pulmonary blood from flowing back into the ventricles and aortic blood into the ventricles.

g ___ papillary muscle

[7] Nipple.

h ___ semilunar valves

[8] A bishop's miter.

i ___ pulmonary semilunar
valve

[9] Tendinous cords.

j ___ aortic semilunar valve

[10] Pulls on the cusps.

[11] Allows blood to enter aorta.

OBJECTIVE 5

G Great Vessles of the Heart (p. 488)

    1 List the five great vessels associated with the heart.

      a

      b

      c

      d

      e

H Blood Supply to the Heart (p. 489)

    1 Check your understanding of the blood supply to the heart by completing this matching exercise.

a ___ coronary arteries

[1] Located in coronary sulcus.

b ___ circumflex branch artery

[2] Where the cardiac veins drain into.

c ___ anastomose

[3] Drains directly into right artrium.

d ___ coronary sinus

[4] To communicate.

e ___ anterior cardiac vein

[5] Open directly into each heart chamber.

f ___ thebesian veins

[6] First branches off the aorta.

III. Physiology of the Heart (p. 490)

    1 What are the two main purposes of the heart?

      a

      b

A Structural and Metabolic Properties of Cardiac Muscle (p. 490)

OBJECTIVE 6

    1 Describe the function of intercalated disks, gap junction, and desmosomes.

    2 Why does heart tissue contain many mitochondria?

    3 What is the role of myoglobin?

B Impulse-Conducting System of the Heart (p. 490)

1 How does the heart beat without any nervous innervation?

OBJECTIVE 7

2 Name, in order, the structures that make up the impulse-conducting system of the heart.

OBJECTIVE 8

C Cardiac Cycle (p. 492)

1 Define the cardiac cycle.

2 Indicate the proper sequence of blood flow through the heart by filling the appropriate letter in the blanks using the following answer code. Start with the right atrium, which is already filled in for you.

   a left ventricle
   b right ventricle
   c left atrium
   d right atrium
   e tricuspid valve
   f bicuspid valve
   g pulmonary valve
   h aortic valve
   i pulmonary circulation
   j systemic circulation

<u>d</u> __ __ __ __ __ __ __ __ __ and back to __d__
1 2 3 4 5 6 7 8 9 10

3 Describe each of the four heart sounds.

   a

   b

   c

   d

OBJECTIVE 9

4 What different factors cause heart sounds?

5 What is a functional heart murmur?

D Nervous Control of the Heart (p. 494)

  1 Indicate how the change listed in Column A affects the item listed in Column B.

Column A	Increases	Decreases	Has no effect on	Column B
a ↓ heart rate	1	2	3	cardiac output
b ↑ stroke volume	1	2	3	cardiac output
c exercise	1	2	3	cardiac output
d ↑ cardiac sympathetic activity	1	2	3	permeability of SA node to K+
e ↑ cardiac sympathetic activity	1	2	3	rate of depolarization of the SA node
f ↑ vagal activity	1	2	3	rate of depolarization of the SA node
g ↑ vagal activity	1	2	3	heart rate
h ↓ venous return	1	2	3	end-diastolic volume
i ↑ end-diastolic volume	1	2	3	stroke volume
j ↓ length of cardiac muscle fiber prior to contraction	1	2	3	stroke volume

  2 What are the two parts of the cardiovascular center?

  a

  b

  3 How do baroreceptors function?

E Endocrine Control of the Heart (p. 495)

  1 Describe how each of the following affects the cardiac output:

  a chemical transmitters

  b adrenal glands

IV. The Effects of Aging on the Heart (p. 495)

  1 What happens as the heart ages?

V. Developmental Anatomy of the Heart (p. 496)
A Early Development of the Heart (p. 496)
    1 Describe how the heart develops.

B Partitioning of the Heart (p. 496)
    1 How is the heart partitioned?

VI. When Things Go Wrong (p. 496)
    1 Describe each of the following disorders:

        a myocardial infarction

        b angina pectoris

        c congestive heart failure

        d pulmonary edema

        e regurgitation

        f stenosis

        g ventricular septal defect

        h interatrial septal defect

        i tetralogy of Fallot

        j cyanosis

        k rheumatic fever

        l pericarditis

        m myocarditis

OBJECTIVE 12

n endocarditis

o circulatory shock

p cardiac tamponade

q cardiac arrhythmias

r atrial fibrillation

s atrial flutter

t AV block

These are terms you should know before proceeding to the post-test.   **MAJOR TERMS**

pulmonary circulation *(p. 479)*

systemic circulation *(p. 479)*

heart *(p. 479)*

apex *(p. 479)*

base *(p. 479)*

pericardium *(p. 479)*

epicardium *(p. 481)*

myocardium *(p. 482)*

endocardium *(p. 482)*

cardiac skeleton *(p. 482)*

right heart *(p. 482)*

left heart *(p. 482)*

septum *(p. 482)*

atrium *(p. 482)*

ventricle *(p. 482)*

coronary sulcus *(p. 483)*

anterior interventricular sulcus *(p. 483)*

posterior interventricular sulcus *(p. 483)*

atrioventricular (AV) sulcus *(p. 483)*

tricuspid valve *(p. 483)*

bicuspid valve *(p. 483)*

semilunar valves *(p. 486)*

pulmonary semilunar valve *(p. 486)*

aortic semilunar valve *(p. 486)*

pulmonary arteries *(p. 489)*

pulmonary veins *(p. 489)*

superior vena cava *(p. 489)*

inferior vena cava *(p. 489)*

aorta *(p. 489)*

left coronary artery *(p. 489)*

right coronary artery *(p. 489)*

coronary sinus *(p. 490)*

pacemaker *(p. 490)*

sinoatrial (SA) node *(p. 490)*

atrioventricular (AV) node *(p. 490)*

atrioventricular bundle (bundle of His) *(p. 491)*

cardiac conduction myofibers (Purkinje fibers) *(p. 491)*

cardiac cycle *(p. 492)*

systole *(p. 492)*

diastole *(p. 492)*

first heart sound (p. 492)

second heart sound *(p. 492)*

third heart sound *(p. 492)*

fourth heart sound *(p. 492)*

heart disease (p. 496)

myocardial infarction *(p. 500)*

congestive heart failure *(p. 500)*

valvular heart disease *(p. 500)*

ventricular septal defect *(p. 500)*

interatrial septal defect *(p. 500)*

tetralogy of Fallot *(p. 500)*

rheumatic fever *(p. 500)*

rheumatic heart disease *(p. 500)*

pericarditis *(p. 501)*

myocarditis *(p. 501)*

endocarditis *(p. 501)*

circulatory shock *(p. 501)*

hypovolemic shock *(p. 501)*

cardiogenic shock *(p. 501)*

cardiac tamponade *(p. 501)*

cardiac arrhythmias *(p. 501)*

angina pectoris *(p. 501)*

CARDIOVASCULAR TECHNOLOGIST AND PERFUSIONIST

The **cardiovascular technologist** performs diagnostic examinations at the request or direction of a physician in one or more of the following three areas: (1) invasive cardiology, (2) noninvasive cardiology, and (3) noninvasive peripheral vascular study. Through subjective sampling and/or recording, the technologist collects data from which the physician can make an anatomic and physiologic diagnosis for each patient. Cardiovascular technologists may provide their services to patients in any medical setting under the supervision of a doctor of medicine or osteopathy. The role of the cardiovascular technologist may include but is not limited to reviewing and/or recording pertinent patient history and supporting clinical data as well as performing appropriate procedures and obtaining a record of anatomical, pathological, and/or physiological data for interpretation by a physician. The technologist learns to exercise discretion and judgment in the performance of cardiovascular diagnostic services.

The procedures performed by the cardiovascular technologist may be found in, but are not limited to, one of the following general settings: (1) invasive cardiovascular laboratories, including cardiac catheterization, blood gas, and electrophysiology laboratories; (2) noninvasive cardiovascular laboratories, including echocardiography, exercise stress test, and electrocardiography laboratories; and (3) noninvasive peripheral vascular studies laboratories, including Doppler ultrasound, thermography, and plethysmography laboratories.

The cardiovascular technologist is qualified by specific technological education to perform these cardiovascular/peripheral vascular diagnostic procedures.

The length of programs varies from 1 to 4 years, depending on student qualifications and which of the three areas of diagnostic evaluation is selected for study. The cardiovascular technologist can choose to master any combination of the three areas.

A **perfusionist** is qualified by academic and clinical education to operate extracorporeal circulation equipment during any medical situation where it is necessary to support or even replace a patient's weak or nonfunctioning circulatory or respiratory system. For example, a perfusionist uses extracorporeal circuits to administer and monitor blood supply, anesthetics, and drugs that a patient must receive during cardiopulmonary bypass surgery. The perfusionist is knowledgeable about the various types of equipment available for extracorporeal circulation and consults with the physician in the selection of appropriate equipment for a patient. A perfusionist may also be responsible for administrative duties such as purchasing supplies and equipment and managing the department and its personnel.

Training programs in perfusion teach the technical aspects of conducting and monitoring a patient's physiological functions with extracorporeal circulation equipment. These programs are generally 1 to 2 years in length, depending on the program's design, objectives, prerequisites, and student qualifications. All programs require a background in the biological sciences; some require specialized background in medical lab technology, respiratory therapy, or nursing.

Based on the terminology contained within this chapter, label the following anterior view of the heart:

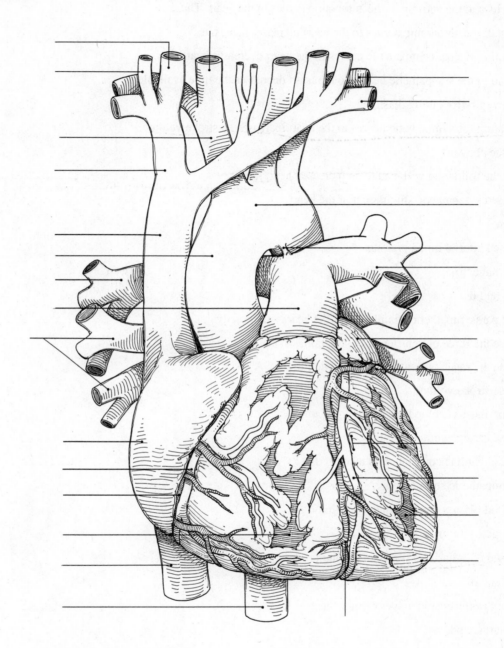

After you have completed the following activities, compare your answers with those given at the end of this manual.

1 The pacemaker of the heart:

   a is in the atrioventricular (AV) node.

   b is the only area of the heart capable of initiating action potentials spontaneously.

   c is slowed by stimulation of the vagus nerve.

   d increases its rate of firing with an increase in body temperature.

   e both **c** and **d**.

2 The activity of the heart depends upon both the inherent properties of the cardiac muscle cells and the activity of the autonomic nerves to the heart. Thus:

   a cutting all the autonomic nerves to the heart increases heart rate.

   b stimulating the parasympathetic nerves to the heart decreases heart rate.

   c stimulating the sympathetic nerves to the heart decreases the time available to fill the ventricles during diastole.

   d stimulating the sympathetic nerves to the heart increases its stroke volume.

   e all of the above.

3 Which of the following will lead to an increased heart rate?

   a decreased sympathetic stimulation of the heart

   b exercise

   c increased arterial blood pressure

   d hemorrhage

   e both **b** and **d**

4 Increased parasympathetic stimulation of the heart will directly:

   a increase the force of ventricular contraction.

   b increase the rate of diastolic depolarization in the SA node.

   c increase stroke volume.

   d decrease diastolic blood pressure.

   e increase cardiac output.

5 The force of cardiac contraction is increased by:

   a parasympathetic stimulation.

   b decreased plasma epinephrine concentration.

   c sympathetic stimulation.

   d increased end-diastolic ventricular volume.

   e both **c** and **d**.

6 The closing of the cuspid valves creates the:

   a first heart sound.

   b second heart sound.

   c third heart sound.

   d fourth heart sound.

   e fifth heart sound.

7 Venous blood is received by the:

   a right atrium.

   b left atrium.

   c right ventricle.

   d left ventricle.

   e none of the above.

8 In a resting individual, the amount of blood pumped by each half of the heart is approximately:

   a 1 L/min.

   b 2 L/min.

   c 3 L/min.

   d 4 L/min.

   e 8 L/min.

9 The inner surface of the heart chambers is known as the:

   a myocardium.

   b endocardium.

   c exocardium.

   d pericardium.

10 The right AV valve is the:

   a mitral.

   b tricuspid.

   c pulmonary.

   d aortic.

   e none of the above.

11 The left coronary artery generally supplies the:

   a right atrium.

   b right ventricle.

   c interventricular septum.

   d all of the above.

   e none of the above.

12 Which of the following is found within the intercalated disks?

   a desmosomes

   b gap junctions

   c both of the above

   d neither of the above

13 The heart receptors for acetylcholine are: _____

    a beta-adrenergic.

    b muscarinic.

    c both of the above.

    d neither of the above.

14 Excess caffeine can cause: _____

    a an action potential in cardiac muscle.

    b extopic foci in cardiac muscle.

    c premature waves of excitation.

    d all of the above.

15 The amount of blood in the ventricle just prior to systole is called the: _____

    a isovolumetric volume.

    b ventricle ejection volume.

    c end-systolic volume.

    d end-diastolic volume

    e both **c** and **d.**

16 Which of the following can alter the heart rate?   _____

    a epinephrine

    b temperature

    c plasma electrolyte concentrations

    d SA node discharge

    e all of the above

17 The left ventricle has to pump more blood than the right ventricle because it has to   _____

pump blood through the whole body, not just through the pulmonary system.

    a true

    b false

18 The cardioregulatory center is located in the medulla of the brainstem.   _____

    a true

    b false

19 The force of contraction of the left ventricle is greater than that of the right   _____

ventricle.

    a true

    b false

20 The AV (atrioventricular) bundle is the only path by which electrical activity occurring in the atria can be conducted to the ventricles.

    a true

    b false

21 The conduction velocity of action potentials is lower in the AV node than in any other portion of the heart muscle.

    a true

    b false

22 Sympathetic stimulation of the SA node decreases the rate at which the membranes of these cells depolarize.

    a true

    b false

23 With each contraction of the ventricles, the entire volume of blood in the ventricle is ejected.

    a true

    b false

The Cardiovascular System: Blood Vessels

20

Prefixes

angio-	vessel
arteri-	an artery
athero-	fatty degeneration
carot-	sleep; stupor
celia-	the abdominal cavity
diastol-	dilation
hepato-	the liver
jugu-	the throat
micro-	small
pulmo-	lungs
systol-	contraction
thrombo-	clot
varic-	enlarged vessel
vaso-	vessel
vol-	volume

Suffixes

-rhage	breaking out
-sclero	hard
-tensin	pressure

Read Chapter 20 in the textbook, focusing on the major objectives. As you progress through it, with your textbook open, answer and/or complete the following. When you complete this exercise, you will have a thorough outline of the chapter.

DEVELOPING YOUR OUTLINE

I. Introduction (p. 507)
   1 What is the general function of blood vessels?

II. Types of Blood Vessels (p. 507)

OBJECTIVE 1

   1 Name the types of vessels a drop of blood can flow through from the aorta back to the right atrium.

   aorta   d _____

   a _____   e _____

   b _____   f _____

   c _____   right atrium

A How Blood Vessels Are Named (p. 507)
   1 Give several examples of how blood vessels are named.

B Arteries (p. 507)

   1 List the three arterial tunics, from the lumen out.

     a

     b

     c

   2 The walls of large arteries are nourished by small blood vessels called the _____.

   3 Explain:

     a pulse

     b tachycardia

     c bradycardia

C Arterioles (p. 508)

   1 Describe the major function of arterioles.

   2 Unlike large arteries, terminal arterioles do not contain:

     a

     b

D Capillaries (p. 509)

   1 Describe the structure of a capillary.

   2 Describe the structure and function of the three different types of capillaries.

     a

     b

     c

OBJECTIVE 2

3 Describe the structure of the walls in the different types of blood vessels.

OBJECTIVE 3

4 What is unique about the capillary blood flow?

5 What vessels constitute the microcirculation?

OBJECTIVE 4

   a

   b

   c

   d

E Venules (p. 511)

   1 List and describe the two types of venules found in the body.

     a

     b

F Veins (p. 511)

   1 Most veins carry deoxygenated blood. However, there are three exceptions. What are they?

     a

     b

     c

   2 Describe how veins differ anatomically from arteries.

III. Circulation of the Blood (p. 512)

OBJECTIVE 5

  A Pulmonary Circulation (p. 512)

   1 The major blood vessels of the pulmonary circulation are the:

     a

     b

c

d

2 What is the main function of the pulmonary circulation?

B Systemic Circulation (p. 512)

OBJECTIVE 6

1 The main vessels of the arterial division of the systemic circulation, from the left ventricle, are the:

a

b

c

d

2 What is the main function of the systemic circulation?

C Portal Systems (p. 512)

OBJECTIVE 7

1 What is unique about a portal system?

2 List the two portal systems found in the human body.

a

b

D Cerebral Circulation (p. 514)

1 What is the function of the cerebral arterial circle (circle of Willis)?

E Cutaneous Circulation (p. 515)

1 What function does the arrangement of blood vessels in the skin serve?

F Skeletal-Muscle Circulation (p. 515)

1 What controls the amount of blood going to skeletal muscles?

G Adaptations in Fetal Circulation (p. 516)

    1 Describe the two main reasons the circulatory system of the fetus differs from that of an adult.

       a

       b

    2 Complete the following table:

Fetal structure	Function	Fate after birth
placenta	Connecting link between mother and fetus.	a _____
b _____	Connects placenta to fetus.	part of afterbirth
umbilical vein	c _____	round ligament
d _____	Joins the inferior vena cava and bypasses the liver.	ligamentum venosum
foramen ovale	e _____	fossa ovalis
f _____	Shunt in the aortic arch; bypasses lungs.	g _____

IV. Major Arteries and Veins (p. 519)

1 Complete the following table on the arterial branches of the aorta:

Artery	Function
a _____	Serves the heart.
b _____	Splits into the right common carotid and right subclavian.
left external carotid	c _____
d _____	Gives off the vertebral artery branch, which serves part of the brain.
e _____	Supply the muscles of the thorax wall.
celiac trunk	A single vessel that has three branches:
f _____	supplies the stomach, the
g _____	supplies the spleen, and the
h _____	supplies the liver.
i _____	Supplies most of the small intestine and the first half of the colon.
inferior mesenteric	j _____
k _____	The final branches of the descending aorta. Each divides into an
l _____	which supplies the pelvic organs, and an
m _____	which enters the thigh where it becomes the femoral artery. At the knee, the femoral artery becomes the
n _____	which then splits into the anterior and
o _____	tibial arteries.

2 Complete the following table of the major veins of the systemic circulation:

Vein	Function
Veins draining into the superior vena cava	
radial and ulnar	Drain the forearm and unite to form the
	a _____
b _____	Empties into the axillary vein.
c _____	Drains the medial aspect of the arm and empties into the brachial vein.
subclavian	d _____
e _____	Drains the posterior part of the head.
internal jugular	f _____
g _____	Drains the thorax.
Veins draining into the inferior vena cava	
anterior and posterior tibial veins	h _____
i _____	The longest veins in the body.
j _____	A single vein that drains the GI tract.
hepatic veins	k _____

V. The Effects of Aging on Blood Vessels (p. 519)

OBJECTIVE 1

1 Describe the following:

   a cerebral arteriosclerosis

   b transient ischemic attack

   c multiple infarct dementia

   d hypertension

   e atherosclerosis

VI. Developmental Anatomy of Major Blood Vessels (p. 538)

1 Describe how blood vessels develop.

VII. When Things Go Wrong (p. 538)

1 Complete the following table on diseases of the blood vessels:

Disease	Symptoms
a _____	Balloonlike dilation of a blood vessel.
arteriosclerosis	b _____
c _____	Leading cause of coronary artery disease.
d _____	A blood clot in a blood vessel.
e _____	A reduced supply of oxygen to heart muscle.
f _____	Commonly called high blood pressure.
patent ductus arteriosus	g _____
stroke	h _____
i _____	Inflammation of a blood vessel.
j _____	Reversal of the great arteries.
k _____	An enlarged saphenous vein.
l _____	Fibrous connective tissue in the liver.

These are the terms you should know before proceeding to the post-test.　**MAJOR TERMS**

arteries *(p. 507)*

capillaries *(p. 507)*

veins *(p. 507)*

aorta *(p. 507)*

pulmonary trunk *(p. 507)*

tunica intima *(p. 507)*

tunica media *(p. 508)*

tunica adventitia *(p. 508)*

vasa vasorum *(p. 508)*

pulse *(p. 508)*

arterioles *(p. 508)*

continuous capillaries *(p. 509)*

fenestrated capillaries *(p. 510)*

discontinuous capillaries *(p. 510)*

sinusoids (vascular sinuses) *(p. 510)*

microcirculation *(p. 510)*

venules *(p. 511)*

pulmonary circulation *(p. 512)*

systemic circulation *(p. 512)*

portal system *(p. 512)*

hypothalamic-hypophyseal portal system *(p. 512)*

hepatic portal system *(p. 513)*

hepatic portal vein *(p. 513)*

splenic veins *(p. 513)*

superior mesenteric veins *(p. 513)*

hepatic veins *(p. 514)*

inferior vena cava (p. 514)

vertebral arteries *(p. 514)*

Based on the terminology contained within this chapter, label the following figures: **LABELING ACTIVITY**

1   An elastic artery.

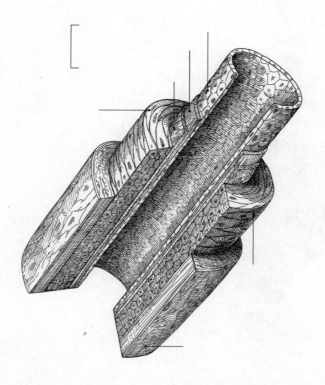

2   Pressure points where the arterial pulse can be felt.

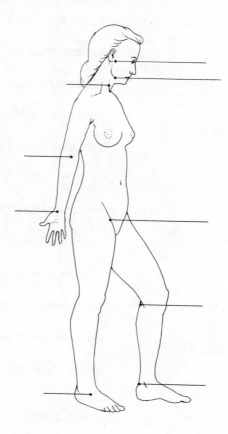

3   A medium-size vein.

4   The systemic circulation.

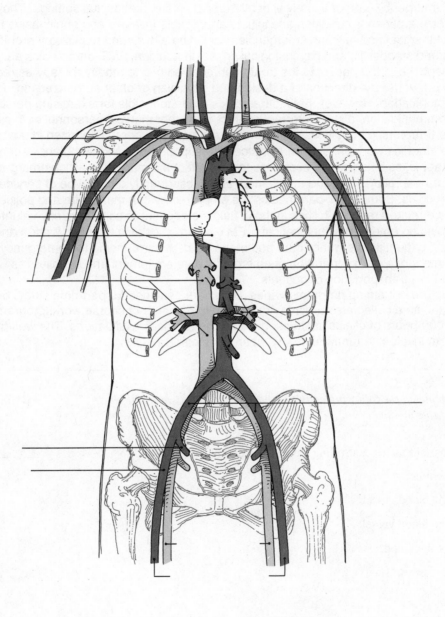

EMERGENCY MEDICAL TECHNICIAN-PARAMEDIC

Emergency medical technicians-paramedics recognize and assess medical emergencies and manage emergency care of acutely ill or injured patients in prehospital settings. They must be able to recognize a patient's condition and initiate appropriate invasive and noninvasive treatments for a variety of surgical and medical emergencies, including airway and respiratory problems, cardiac dysrhythmias and standstills, and psychological crises. In addition, EMT-paramedics are able to assess the response of the patient to the treatment received and to modify therapy as required. The paramedics work under the direction of a designated physician or other authorized individual, often by radio communication. However, EMT-paramedics are responsible for exercising personal judgment when communication failures interrupt contact with other medical personnel or in cases of immediate life-threatening conditions when emergency care has been authorized in advance.

EMT-paramedics work primarily in advanced life-support units and ambulance services. Variations in geographic, sociologic, and economic factors affect the manner in which emergency medical service systems are recognized and maintained, and subsequently on the type of services provided by EMT-paramedics. Some EMT-paramedics are employed by community fire and police departments and have related responsibilities in those fields; others serve as community volunteers. In addition to providing emergency medical care, EMT-paramedics also maintain written and verbal records of each patient's emergency care and of the incident surrounding the emergency. They also direct the maintenance and preparation of emergency care equipment and supplies, as well as direct and coordinate the transportation of patients.

The structure of training programs varies. Some are designed for part-time study, others are organized as full-time collegiate curricula. In addition to 60 hours of course work, programs require from 600 to 1000 hours of clinical instruction in emergency medical situations. The well-trained EMT-paramedic is an invaluable member of any emergency medical team.

After you have completed the following activities, compare your answers with those given at the end of this manual.

POST-TEST

1 The most important factor controlling the resistance to blood flow is: _____

  a body temperature.

  b radius of the blood vessel.

  c length of the blood vessel.

  d viscosity of the blood.

  e the blood pressure.

2 Matching.

  a ___ peroneal artery   [1] Forearm and wrist.   _____

  b ___ subscapular artery   [2] Chest and shoulder.   _____

  c ___ occipital artery   [3] Knee joint.   _____

  d ___ common interosseous   [4] Leg.   _____
       artery

  e ___ superior genicular artery   [5] Cerebral meninges.   _____

3 Large arteries have all of the following tissue layers *except* the:

   a nerve-conducting layer.

   b connective tissue layer.

   c elastic elements.

   d muscle fibers.

   e both **a** and **b**.

4 The baroreceptors that are sensitive to arterial blood pressure changes are located in the _____ and _____ .

5 _____ (high blood pressure) is defined as chronically increased arterial pressure.

6 Insufficient coronary blood flow can lead to a myocardial _____ .

7 A _____ is another name for a cerebral occlusion.

8 The _____ artery supplies blood to the left arm.

9 The blood supply to the brain is provided by the _____ .

10 The largest artery is the aortic trunk.

   a true

   b false

11 The site at which a capillary leaves a capillary bed is protected by a ring of smooth muscle called the:

   a venule valve.

   b arteriolar sphincter.

   c terminal sphincter.

   d precapillary sphincter.

12 Which of the following is *not* a portal system of the body?

   a adrenal portal system

   b hepatic portal system

   c renal portal system

   d hypophyseal portal system

   e both **a** and **c**

13 The entire system of blood vessels has one structural component in common: a smooth, low-friction lining of _____ .

14 The thickest portion of an artery is the _____ , whereas the thickest layer in a vein is the _____ .

15 True or False.

    i Arterial walls are free of nerves.   _____

    a true

    b false

    ii The blood velocity is faster in arterioles than in the aortic artery.   _____

    a true

    b false

    iii Venous return affects atrial pressure.   _____

    a true

    b false

16 In humans, blood pressure can be measured with an instrument called a   _____

    _____.

17 The _____ safeguards the supply of blood to all parts of the brain.   _____

18 The _____ , formed by the union of the right and left common iliac   _____

    veins, returns blood to the heart.

19 All veins have valves.   _____

    a true

    b false

20 Veins can accommodate larger volumes of blood with little increase of internal   _____

    pressure because:

    a veins have valves.

    b veins are distensible.

    c veins are thinner than arteries.

    d both **b** and **c**.

    e both **c** and **d**.

21 The two vertebral arteries unite to form the: _____

    a cerebral arterial circle (circle of Willis).

    b anterior cerebral arteries.

    c internal carotid artery.

    d posterior cerebral artery.

    e splenic artery.

22 The round ligament of the liver is formed by the: _____

    a umbilical vein.

    b ductus venosus.

    c hepatic portal vein.

    d umbilical arteries.

    e cerebral arterial circle (circle of Willis).

23 If a person has a blood pressure reading of 130/80, this individual's pulse pressure would be:

    a 210 mm Hg.

    b 120 mm Hg.

    c 60 mm Hg.

    d 50 mm Hg.

    e 20 mm Hg.

_____

24 Matching.

    a ___ ductus venosus   **[1]** A shunt between the atria that bypasses the pulmonary blood vessels.   _____

    b ___ ductus arteriosus   **[2]** Transports blood from the fetus to the placenta.   _____

    c ___ umbilical arteries   **[3]** Transports oxygenated blood directly into the inferior vena cava.   _____

    d ___ umbilical vein   **[4]** A shunt between the pulmonary trunk and aortic arch that bypasses the pulmonary cells.   _____

    e ___ foramen ovale   **[5]** Transports nutrient-rich oxygenated blood from the placenta to the fetus.   _____

25 Matching.

    a ___ left carotid artery   **[1]** Lower extremities.   _____

    b ___ phrenic artery   **[2]** Heart.   _____

    c ___ left coronary artery   **[3]** Brain.   _____

    d ___ inferior mesenteric artery   **[4]** Diaphragm.   _____

    e ___ external iliac arteries   **[5]** Rectum and anus.   _____

The Lymphatic System

21

Prefixes

anti-	against
glutin-	glue
immun-	free
reticulo-	a network
trab-	beam; timber

Suffixes

-gen	to bear
-phage	to eat

Read Chapter 21 in the textbook, focusing on the major objectives. As you progress through it, with your textbook open, answer and/or complete the following. When you complete this exercise, you will have a thorough outline of the chapter.

DEVELOPING YOUR OUTLINE

I. Introduction (p. 545)

OBJECTIVE 1

1 Describe the four major functions of the lymphatic system.

a

b

c

d

II. The Lymphatic System (p. 545)

1 Describe the component parts of the lymphatic system.

a

b

c

d

e

f

A Lymph (p. 545)

   1 Describe how lymph is produced.

   2 Describe the composition of lymph.

   3 Complete the following table on cells of the lymph fluid:

Cell	Function/description
lymphocytes	**a** _____
b _____	Make up the reticuloendothelial system.
c _____	Attached tissue macrophages; walling-off.
B lymphocyte	**d** _____
T lymphocytes	**e** _____

B Lymphatic Capillaries and Other Vessels (p. 548)

   1 Contrast the structure of blood and lymphatic capillaries.

   2 Contrast the structure of veins and lymphatics.

C Circulation of Lymph (p. 549)

   1 Describe the three forces responsible for lymph movement.

     a

     b

     c

   2 Name the areas of the body drained by the following lymphatics:

     a right lymphatic duct _____ .

     b left lymphatic duct _____ .

     c cisterna chyli _____ .

D Lymph Nodes (p. 550)

   1 Where in the body are the greatest concentrations of lymph nodes found?

      a

      b

      c

      d

      e

   2 Describe the function of lymph nodes.

   3 Complete the following table on the structure of a typical lymph node:

Anatomical part	Function
a _____	fibrous connective tissue covering
b _____	divide the lymph node into compartments
c _____	outer portion
d _____	dense cluster of lymphocytes
e _____	center for lymphocyte production
f _____	vessel leaving the lymph node

E Tonsils (p. 551)

   1 Name and describe the location of the body's three tonsils.

      a

      b

      c

   2 What is the major function of the tonsils?

F Spleen (p. 551)
  1 Describe the functions of the spleen.

  a

  b

  c

  d

  e

  2 Describe the following structures of the spleen:
  a white pulp

  b red pulp

G Thymus Gland (p. 552)
  1 Briefly describe the structure, function, and location of the thymus gland.

H Aggregated Lymph Nodules (Peyer's Patches) (p. 553)
  1 Where would you look for aggregated lymph nodules?

  2 What are their functions?

III. Developmental Anatomy of the Lymphatic System (p. 553)
  1 Describe how the lymphatic system develops.

IV. When Things Go Wrong (p. 553)
  A Acquired Immune Deficiency Syndrome (AIDS) (p. 554)
    1 What causes AIDS?

    2 What cells are affected in AIDS?

    3 How can AIDS be treated?

B Briefly describe each of the following disorders:

    1 Hodgkin's disease

    2 Infectious mononucleosis

These are terms you should know before proceeding to the post-test.   **MAJOR TERMS**

lymphatic system *(p. 545)*

interstitial fluid *(p. 545)*

lymph *(p. 545)*

monocytes *(p. 545)*

macrophages *(p. 545)*

histocytes *(p. 545)*

B cells *(p. 545)*

T cells *(p. 545)*

lymphatic capillaries *(p. 548)*

lacteals *(p. 548)*

lymphatics *(p. 548)*

right lymphatic duct *(p. 549)*

thoracic duct *(p. 549)*

lymph nodes *(p. 550)*

tonsils *(p. 551)*

spleen *(p. 551)*

thymus gland *(p. 552)*

aggregated lymph nodules (Peyer's patches) *(p. 553)*

acquired immune deficiency syndrome (AIDS) *(p. 554)*

human immunodeficiency virus (HIV) *(p. 554)*

retrovirus *(p. 555)*

Hodgkin's disease *(p. 556)*

infectious mononucleosis *(p. 556)*

MEDICAL ASSISTANT

The **medical assistant** is a professional, multiskilled person dedicated to assisting in patient care management. This practitioner performs administrative and clinical duties and may manage emergency situations, facilities, and/or personnel. Competence in the field also requires that a medical assistant display professionalism, communicate effectively, and provide instruction to patients. Medical assistants assist physicians in their offices or other medical settings, performing administrative and/or clinical duties delegated in relation to the degree of training and in accordance with respective state laws governing such actions and activities. Medical assistants have a wide range of duties.

Their business administrative duties include scheduling and receiving patients; obtaining patients' data; maintaining medical records; typing and medical transcription; handling telephone calls, correspondence, reports, and manuscripts; and assuming responsibility for office care, insurance matters, office accounts, and fees and collections.

Their clinical duties may include preparing the patient for examination, obtaining vital signs, taking medical histories, assisting with examinations and treatments, performing routine office laboratory procedures and electrocardiograms, sterilizing instruments and equipment for office procedures, and instructing patients in preparation for x-ray and laboratory examinations.

Both administrative and clinical duties involve purchasing and maintaining supplies and equipment. A medical assistant who is sufficiently qualified by education and/or experience may be responsible for personnel and office management.

Associate degree programs in medical assistance are two years long; certificate or diploma programs are one year long.

Based on terminology contained within this chapter, label the following components of the lymphatic system:

LABELING ACTIVITY

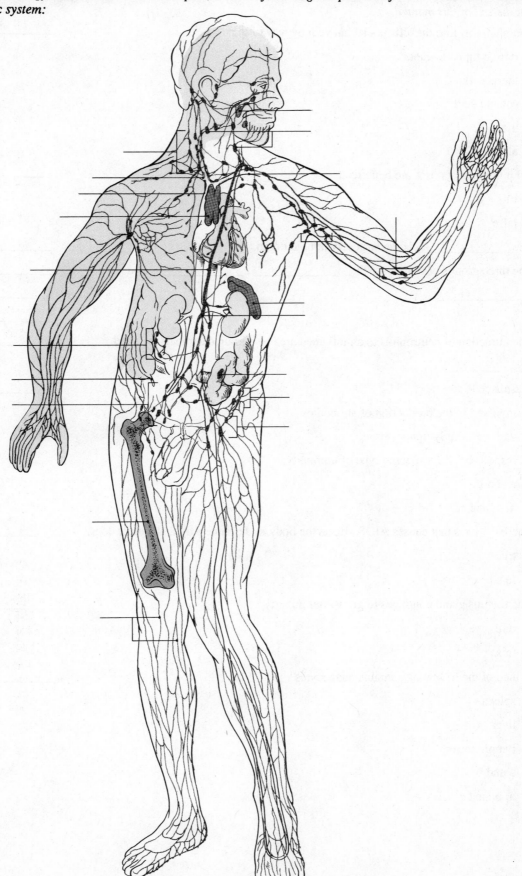

After you have completed the following activities, compare your answers with those given at the end of this manual.

1 Lymph drains into the left subclavian vein by way of the:

    a right lymphatic duct.

    b thoracic duct.

    c splenic duct.

    d a and **b,**

    e a, **b,** and **c.**

       _____

2 All immune responses are beneficial.

    a true

    b false

       _____

3 The _____ is the largest lymphoid organ.

       _____

4 The three groups of tonsils are the _____ , _____ and

    _____ .

       _____

       _____

       _____

5 The attraction of neutrophils to an inflamed area by certain chemicals is called

    _____ .

       _____

6 Plasma cells are:

    a involved in the production of antibodies.

    b derived from lymphocytes.

    c responsible for a specific type of immunity.

    d a and **b.**

    e a, **b,** and **c.**

       _____

7 The HIV virus that causes AIDS affects the body in the same way that cancer does.

    a true

    b false

       _____

8 The thymus gland continues to grow after puberty.

    a true

    b false

       _____

9 Which of the following contains phagocytes?

    a spleen

    b liver

    c lymph nodes

    d a and **b**

    e a, **b,** and **c**

       _____

10 In B lymphocytes, the letter B originates from: _____

    a an immunologist named Bueller.

    b the lymphocyte bursa.

    c the bursa of Fabricus found in chickens.

    d **a** and **b.**

    e **a, b,** and **c.**

The Respiratory System

22

Prefixes

alveol-	cavity; small hollow
bronch-	air passage
crico-	a ring
dia-	across
epi-	upon
naso-	nose
oro-	mouth
pleur-	the side; rib
pneumo-	lung; breath
pulmo-	a lung
spiro-	breathe
tax-	arrangement
thyro-	a shield

Suffixes

-itis	inflammation
-phragm	partition

Read Chapter 22 in the textbook, focusing on the major objectives. As you progress through it, with your textbook open, answer and/or complete the following. When you complete this exercise, you will have a thorough outline of the chapter.

DEVELOPING YOUR OUTLINE

I. Introduction (p. 560)

1 Give several meanings for the word *respiration*.

2 What is the difference between internal and external respiration?

3 List some of the functions of the respiratory system.

a

b

c

d

e

OBJECTIVE 1

II. Respiratory Tract (p. 560)
    1 List the structures of the respiratory tract.

OBJECTIVE 2

  A Nose (p. 560)
    1 Describe the structures that carry out the following:

      a separates the external nares _____

      b serve as air passages _____

      c forms the floor of the nasal cavity _____

      d contribute moisture to the nasal cavity _____

      e involved in the sense of smell _____

OBJECTIVE 3

  B Pharynx (p. 563)
    1 List and describe the three parts of the pharynx and the function of each.

      a

      b

      c

OBJECTIVE 4

  C Larynx (p. 564)
    1 Describe the following structures of the larynx:

      a thyroid cartilage

      b cricoid cartilage

      c epiglottis

      d glottis

    2 Describe how the larynx produces sound.

OBJECTIVE 5

  D Trachea (p. 564)
    1 Describe the function(s) of the following parts of the trachea:

      a cartilaginous rings

b ciliated columnar epithelium

2 What is the relationship between the trachea and respiratory tract?

E Lungs: The Respiratory Tree (p. 567)

OBJECTIVE 6

1 Complete the following table on the histology of the respiratory tree:

Structure	Tissue type
bronchi	a _____
bronchioles	b _____
alveolar ducts	c _____
alveoli	d _____

OBJECTIVE 7

2 Describe the functions of the following structures associated with an alveolus:

a basement membrane

b alveolar sac

c type I cells

d septal cells

e macrophages

f capillary network

3 What is the function of surfactant?

F Lungs: Lobes and Pleurae (p. 570)

OBJECTIVE 8

   1 Complete the following exercise on the lungs:

The lungs fill the thoracic cavity except for a midventral region, called

the **a** _____ . The top of the lung is pointed and called the

b _____ . The three lobes of the right lung are called

c _____ , **d** _____ , and **e** _____ . In

the left lung, the heart fits into the **f** _____ . Within the lungs,

the bronchopulmonary segments contain smaller subdivisions called

g _____ . The lungs are covered by an outer surface mem-

brane called the **h** _____ and one that lines the chest cavity

called the **i** _____ . The space between the two pleurae is

called the **j** _____ .

   2 Describe the three functions associated with the pleurae.

     a

     b

     c

G Nerve and Blood Supply (p. 572)

   1 Bronchodilation of the tracheobronchial tree can be caused by:

     a

     b

   2 Describe the blood supply of the respiratory tree.

III. Mechanics of Breathing (p. 573)

OBJECTIVE 9

   1 How does air move into and out of the lungs?

A Muscular Control of Breathing (p. 573)

   1 What muscles are involved in inspiration? What does each do?

     a

     b

2 What muscles are involved in forced expiration?

a

b

IV. Other Activities of the Respiratory System (p. 574)

1 List and describe several other activities of the respiratory system in addition to gas exchange.

a

b

c

d

e

f

g

h

i

V. The Effects of Aging on the Respiratory System (p. 575)

1 What are some effects of aging on the respiratory system?

VI. Developmental Anatomy of the Respiratory Tree (p. 576)

1 Describe how a lung develops.

VII. When Things Go Wrong (p. 577)

1 Discuss the following disorders or accidents that affect the respiratory system:

a rhinitis

b laryngitis

c asthma

d bronchial asthma

e hay fever

f emphysema

g pneumonia

h tuberculosis

i lung cancer

j pleurisy

k drowning

l cyanosis

m hyaline membrane disease

n sudden infant death syndrome

These are terms you should know before proceeding to the post-test.   **MAJOR TERMS**

respiratory system *(p. 560)*   ventilation (breathing) *(p. 560)*

cellular respiration *(p. 560)*   inspiration *(p. 560)*

external respiration *(p. 560)*   expiration *(p. 560)*

internal respiration *(p. 560)*   nasal bones *(p. 560)* external nares *(p. 560)*

nasal septum *(p. 560)*

nasal cavity *(p. 560)*

nasolacrimal duct *(p. 563)*

pharynx *(p. 563)*

nasopharynx *(p. 563)*

oropharynx *(p. 563)*

laryngopharynx *(p. 563)*

tonsils *(p. 564)*

larynx (voice box) *(p. 564)*

epiglottis *(p. 564)*

vocal cords *(p. 564)*

trachea *(p. 564)*

bronchi *(p. 564)*

bronchioles *(p. 567)*

terminal bronchioles *(p. 567)*

alveoli *(p. 567)*

surface tension *(p. 567)*

lungs *(p. 570)*

pleural cavity *(p. 570)*

diaphragm *(p. 573)*

external intercostal muscles *(p. 573)*

internal intercostal muscles *(p. 574)*

coughing reflex *(p. 574)*

sneeze *(p. 574)*

Heimlich maneuver *(p. 575)*

hiccup (p. *575)*

snoring (p. *575)*

rhinitis *(p. 577)*

laryngitis *(p. 577)*

asthma *(p. 577)*

hay fever *(p. 577)*

emphysema *(p. 577)*

pneumonia *(p. 578)*

tuberculosis *(p. 578)*

cancer of the lungs *(p. 578)*

pleurisy *(p. 578)*

drowning *(p. 578)*

cyanosis *(p. 578)*

hyaline membrane disease *(p. 579)*

sudden infant death syndrome *(p. 579)*

RESPIRATORY THERAPIST

The **respiratory therapist** applies his or her knowledge of respiratory function to clinical problems of respiratory care. The respiratory therapist is qualified to assume primary responsibility for all respiratory care modalities, including the supervision of respiratory therapy technicians. The respiratory therapist may be required to exercise considerable independent, clinical judgment in the respiratory care of patients under the supervision of a physician.

Knowledge and skills for performing these functions are achieved through formal programs of didactic, laboratory, and clinical preparation. A two-year course of study leading to an associate degree is the minimum training required; in some instances, a baccalaureate degree is required.

Based on the terminology contained within this chapter, label the following figures: **LABELING ACTIVITY**

1   The respiratory system.

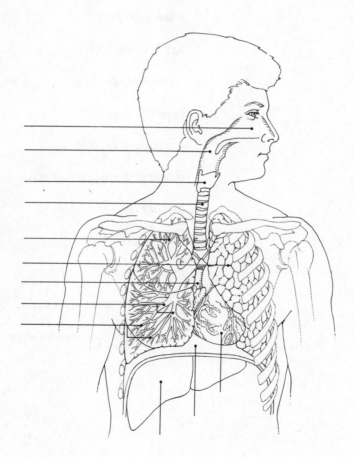

2   Larynx and trachea.

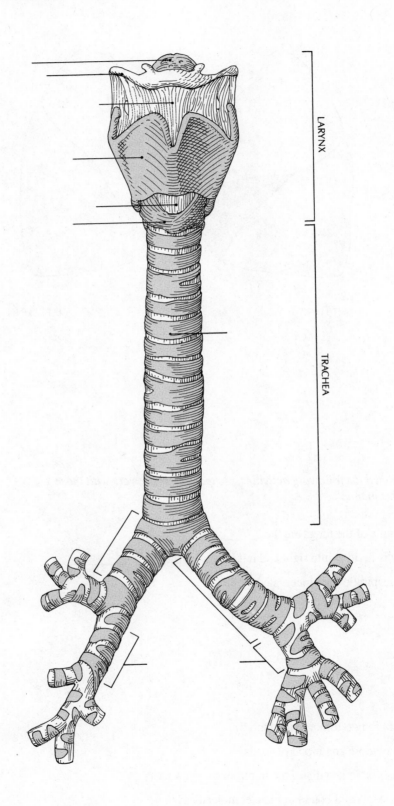

3   Lobes of the lungs.

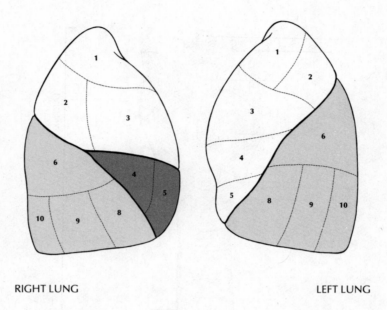

RIGHT LUNG   LEFT LUNG

After you have completed the following activities, compare your answers with those given at the end of this manual.   **POST-TEST**

1 The functional units of the lungs are the _____ .

2 The nasal cavity is divided into right and left halves by the _____ .

3 The muscles of respiration are innervated by the:

   a vagus nerve.

   b spinal accessory nerve.

   c phrenic nerves.

   d a and b.

   e a, b, and c.

4 Oxygen is present in two forms in the blood:

   a bound to hemoglobin and blood platelets.

   b bound to hemoglobin and dissolved in plasma.

   c dissolved in plasma and bound to platelet molecules.

   d none of the above.

5 The neural apparatus for regulating breathing is located in the hypothalamus.   _____

   a true

   b false

6 A hiccup can be defined as a spasmodic contraction of the diaphragm.   _____

   a true

   b false

7 C-shaped cartilaginous rings are present in the _____ of the respiratory   _____
   system.

8 Matching.

   a ___ parietal pleura   [1] H_2CO_3.   _____

   b ___ larynx   [2] Outer layer of thoracic lining.   _____

   c ___ pneumonia   [3] Contains the vocal cords.   _____

   d ___ emphysema   [4] Fluid accumulation in alveoli.   _____

   e ___ carbonic acid   [5] Loss of alveolar elasticity.   _____

9 Which of the following muscle(s) is/are *not* used in taking air into the lungs?   _____

   a external intercostals

   b internal intercostals

   c scalenes

   d diaphragm

   e both **a** and **b**

10 The movement and capture of foreign particles traveling from the mouth to the   _____
    lungs results from all of the following *except*:

   a mucus.

   b goblet cells.

   c cilia.

   d squamous epithelium.

   e both **a** and **b**.

11 Which of the following is *not* an obstructive lung disease?   _____

   a asthma

   b pneumothorax

   c emphysema

   d pneumonia

   e both **c** and **d**

12 The lungs are located within the pleural cavity.   _____

   a true

   b false

13 The principal supporting structure within the bronchial walls is _____ .   _____

14 The exchange of air between the atmosphere and the alveoli is called   _____

_____ .

15 The exchange of gas between the cells of the body and the external environment is   _____

_____ .

16 The medullary inspiratory neurons receive inputs from several major sources.   _____

Which of the following illustrates these sources?

   a reciprocal connections with the medullary expiratory neurons

   b connections with the pons

   c stretch receptors in the lungs

   d a and b

   e a, b, and c

17 True or False.

   i The cilia of the lungs provide a defense against infection.   _____

   a true

   b false

   ii The alveoli are filled with intrapleural fluid.   _____

   a true

   b false

   iii The cilia help move mucus along the air passage toward the larynx.   _____

   a true

   b false

   iv The cilia are found in the pleura near the alveoli.   _____

   a true

   b false

   v The cilia are hairlike projections on the epithelial linings of the pharynx.   _____

   a true

   b false

The Digestive System

23

Prefixes

aliment-	food
api-	the tip
bili-	bile
bucc-	the cheek
cari-	decay
cec-	blindness
chyl-	juice
entero-	intestine
frenul-	a restraint
gastr-	stomach
gingiv-	the gums
hepat-	liver
hiat-	an opening
lys-	loosening
mastic-	chew
sigm-	Greek Σ; Roman S
sin-	cavity; recesss
stal-	contract
sulc-	furrow; groove
vill-	hairy

Suffixes

-flux	flow
-in	a protein
-ose	a carbohydrate

Read Chapter 23 in the textbook, focusing on the major objectives. As you progress through it, with your textbook open, answer and/or complete the following. When you complete this exercise, you will have a thorough outline of the chapter.

DEVELOPING YOUR OUTLINE

I. Introduction (p. 582)

1 List the five basic activities involved in digestion and absorption.

a

b

c

d

e

2 Differentiate between digestion and absorption.

OBJECTIVE 1

II. Introduction to the Digestive System (p. 583)

1 Contrast chemical and mechanical digestion.

2 List the organs of the alimentary canal.

OBJECTIVE 2

   a

   b

   c

   d

   e

   f

   g

   h

   i

3 List the associated structures of the digestive system.

   a

   b

   c

   d

   e

   f

g

h

i

A Basic Functions (p. 583)
   1 List a basic function of the following parts of the digestive system:
   a mouth

   b pharynx

   c esophagus

   d stomach

   e small intestine

   f large intestine

   g associated structures

B Tissue Structure (p. 583)   OBJECTIVE 3
   1 List and briefly describe the four layers of tissue in the wall of the digestive tract.
   a

   b

   c

   d

III. Mouth (p. 583)

    1 What are two other anatomical names for the mouth?

      a

      b

    2 What structures make up the oral cavity?

      a

      b

    3 What one word describes the mechanical action that takes place in the mouth? _____

A Lips and Cheeks (p. 587)

    1 List one function of the lips.

    2 What is the function of the cheeks?

B Teeth and Gums (p. 587)

    1 A normal child has _____ deciduous teeth.

    2 A normal adult has _____ permanent teeth.

    3 In an adult, each jaw has _____ incisors, _____ canines, _____ premolars, and _____ .

    4 Describe the three major parts of each tooth.

      a

      b

      c

    5 What is the function of the periodontal ligaments?

    6 Describe each of the following parts of a tooth:

      a dentine

      b enamel

c cement

d pulp

C Tongue (p. 588)

   1 Describe the two major functions of the tongue.

     a

     b

   2 Which cranial nerve innervates the tongue?

D Palate (p. 588)

   1 Describe the two sections of the palate.

     a

     b

   2 Describe the function of the hard palate.

   3 What is the function of the soft palate?

OBJECTIVE 6

E Salivary Glands (p. 590)

   1 List the three pairs of salivary glands.

     a

     b

     c

   2 Describe several functions of saliva.

   3 What is the specific function of salivary amylase?

4 List several factors that stimulate salivary secretion.

5 Describe the control of salivary secretion.

IV. Pharynx and Esophagus (p. 591)
    A Pharynx (p. 591)
        1 Describe the three anatomical parts of the pharynx.

            a

            b

            c

    B Esophagus (p. 591)
        1 From the inside out, list and describe the tissue layers of the esophagus.

            a

            b

            c

        2 What are the two sphincters associated with the esophagus?

            a

            b

        3 The esophagus connects the _____ to the

           _____ .

        4 When hydrochloric acid from the stomach enters the esophagus, this is
           commonly called _____ .

C Swallowing (Deglutition) (p. 592)

    1 Describe the role each of the following structures plays in the swallowing process:

        a tongue

        b soft palate

        c epiglottis

        d pharyngeal constrictor muscle

    2 Describe the three stages of swallowing, along with the structures involved with the processes.

> **OBJECTIVE 7**

        a

        b

        c

V. Abdominal Cavity and Peritoneum (p. 593)

> **OBJECTIVE 8**

    1 Contrast the parietal and visceral peritoneums. What is the space between these two membranes called? What is found within this space?

    2 List the abdominal organs that are retroperitoneal.

        a

        b

        c

        d

        e

        f

3 Which organs are intraperitoneal?

   a

   b

   c

   d

4 Match the following peritoneal extensions:

a ___ greater omentum   **[1]** Suspends the stomach and duodenum from the liver.

b ___ lesser omentum   **[2]** Attaches the liver to the abdominal wall.

c ___ mesentery   **[3]** Attaches the large intestine to the posterior abdominal wall.

d ___ falciform ligament   **[4]** Covers the small intestine.

e ___ mesocolon   **[5]** Contains many blood and lymph vessels.

VI. Stomach (p. 595)

  A Anatomy of the Stomach (p. 595)

    1 Describe the location of the stomach.

    2 Name the two orifices of the stomach.

      a

      b

    3 Describe the four major regions of the stomach.   OBJECTIVE 9

      a

      b

      c

      d

4 Name the three layers of the muscularis externa.

OBJECTIVE 10

a

b

c

5 Complete the following table on the types of stomach cells:

OBJECTIVE 11

Type of cell	Description	Function
a _____	line stomach lumen	secrete alkaline mucus
neck mucous cells	line glands	b _____
c _____	found in fundic glands	secrete HCl
chief cells	d _____	secrete pepsinogen
enteroendocrine cells	found at base of gastric glands	e _____

VII. Small Intestine (p. 598)

OBJECTIVE 12

A Anatomy of the Small Intestine (p. 598)

1 Name the four parts of the duodenum.

a

b

c

d

2 The ileum joins the cecum at the _____ .

3 Describe the three unique features of the small intestine that enhance the digestion and absorption processes.

OBJECTIVE 13

a

b

c

4 What is the function of submucosal glands (Brunner's glands)? Aggregated Lymph nodules (Peyer's patches)?

5 Complete the following table on the cell types found in the small intestine:

Cells	Function
columnar	a _____
b _____	source of intestinal cells
c _____	secrete mucus
Paneth	d _____
enteroendocrine	e _____

B Functions of the Small Intestine (p. 600)

    1 Describe the two main types of muscular activity in the small intestine.

      a

      b

C Absorption from the Small Intestine (p. 602)

OBJECTIVE 14

    1 Monosaccharides are readily absorbed through the _____ on the border of columnar absorptive cells. An _____ mechanism is used for the absorption of amino acids. Fats enter the small intestine in the form of _____ . These are acted upon by pancreatic _____ , which breaks the droplets down into free fatty acids, glycerol, and _____ . Bile salts from the liver _____ the liver droplets. Water-soluble particles called _____ are thus formed. After absorption, protein-coated droplets called _____ move into the _____ of each villus.

    2 _____ is needed for absorption of vitamin B_{12}.

VIII. Pancreas As a Digestive Organ (p. 604)

OBJECTIVE 15

    1 Describe the three anatomical parts of the pancreas.

      a

b

c

2 Describe three ducts associated with the pancreas.
a

b

c

IX. Liver As a Digestive Organ (p. 604)

OBJECTIVE 16

A Anatomy of the Liver (p. 604)

1 The liver is covered by connective tissue called _____ . It is held in place by the _____ . The liver is divided into two main lobes by the _____ . The fibrous cord that is the remnant of the left fetal umbilical vein is the _____ . The right lobe of the liver is subdivided into a small _____ lobe and a small _____ artery supplies the liver with blood and the _____ vein drains venous blood from the GI tract. Within the liver, _____ are formed by the bile capillaries. The _____ regulates the outflow of bile into the duodenum. The functional units of the liver are the _____ . Liver cells ar called _____ . Fixed macrophages within the liver are called

_____ .

2 Describe the microscopic anatomy of the liver.

B Functions of the Liver (p. 606)

1 Describe several metabolic functions of the liver.

X. Gallbladder and Biliary System (p. 606)

1 Anatomically, the biliary system consists of:

a

b

c

d

2 What role does the gallbladder play in the digestive process?

OBJECTIVE 17

XI. Large Intestine (p. 608)

OBJECTIVE 18

A Anatomy of the Large Intestine (p. 609)

1 List the anatomical parts of the large intestine.

a

b

c

d

2 How does the large intestine differ physically from the small intestine?

a

b

c

B Rectum, Anal Canal, and Anus (p. 609)

1 What two muscles hold the anal canal closed?

a

b

C Functions of the Large Intestine (p. 611)

1 Describe the primary functions of the large intestine.

2 How is the nervous system involved with the following movements of the large intestine?

a haustral contractions

b mass movements

XII. The Effects of Aging on the Digestive System (p. 612)

1 List some things that happen to the digestive system as one ages.

XIII. Developmental Anatomy of the Digestive System (p. 612)

1 Describe how the digestive system develops.

XIV. When Things Go Wrong (p. 613)

1 Briefly describe each of the following disorders:

a anorexia nervosa

b bulimia

c cholelithiasis

d cirrhosis

e colon-rectum cancer

f constipation

g diarrhea

h Crohn's disease

i diverticulosis

j diverticulitis

k food poisoning

l hepatitis

m hernia

n jaundice

o peptic ulcers

p peritonitis

q tooth decay (dental caries)

r vomiting

These are terms you should know before proceeding to the post-test.   **MAJOR TERMS**

digestion *(p. 583)*	root *(p. 588)*
absorption *(p. 583)*	crown *(p. 588)*
ingestion *(p. 583)*	neck *(p. 588)*
peristalsis *(p. 583)*	dentine *(p. 588)*
defecation *(p. 583)*	enamel *(p. 588)*
chemical digestion *(p. 583)*	cement *(p. 588)*
mechanical digestion *(p. 583)*	pulp *(p. 588)*
digestive tract (alimentary canal) *(p. 583)*	gum (gingiva) *(p. 588)*
mucosa *(p. 583)*	tongue *(p. 588)*
submucosa *(p. 583)*	palate *(p. 588)*
muscularis externa *(p. 583)*	hard palate *(p. 590)*
serosa *(p. 583)*	soft palate *(p. 590)*
mouth (oral cavity) *(p. 583)*	salivary glands *(p. 590)*
lips *(p. 587)*	saliva *(p. 590)*
cheeks *(p. 587)*	parotid glands *(p. 590)*
deciduous teeth *(p. 587)*	submandibular glands *(p. 590)*
incisors *(p. 587)*	sublingual glands *(p. 590)*
canines *(p. 587)*	pharynx *(p. 591)*
premolars *(p. 587)*	esophagus *(p. 591)*
molars *(p. 587)*	bolus *(p. 592)*
wisdom teeth *(p. 587)*	deglutition *(p. 592)*

abdominal cavity *(p. 593)*

abdominopelvic cavity *(p. 593)*

peritoneum *(p. 593)*

greater omentum *(p. 594)*

lesser omentum *(p. 594)*

stomach *(p. 595)*

cardiac (esophageal) orifice *(p. 595)*

pyloric orifice (pyloris) *(p. 595)*

cardiac region *(p. 595)*

fundus *(p. 595)*

body *(p. 595)*

pyloric region *(p. 595)*

rugae *(p. 597)*

chyme *(p. 598)*

gastric juice *(p. 598)*

small intestine *(p. 598)*

duodenum *(p. 598)*

jejunum *(p. 598)*

ileum *(p. 598)*

plicae circulares *(p. 600)*

villi *(p. 600)*

intestinal glands (crypts of Lieberkühn) *(p. 600)*

lacteal *(p. 600)*

columnar absorptive cells *(p. 600)*

undifferentiated cells *(p. 600)*

mucous goblet cells *(p. 600)*

Paneth cells *(p. 600)*

enteroendocrine cells *(p. 600)*

segmenting contractions *(p. 600)*

peristaltic contractions *(p. 600)*

micelles *(p. 602)*

chylomicrons *(p. 602)*

pancreas *(p. 604)*

acini *(p. 604)*

liver *(p. 604)*

porta hepatis *(p. 604)*

bile canaliculi *(p. 605)*

bile capillaries *(p. 605)*

hepatic ducts *(p. 605)*

cystic duct *(p. 605)*

common bile duct *(p. 605)*

lobules *(p. 605)*

hepatic cells *(p. 606)*

sinusoids *(p. 606)*

bile *(p. 606)*

gallbladder *(p. 606)*

biliary system *(p. 608)*

feces *(p. 608)*

large intestine *(p. 609)*

cecum *(p. 609)*

vermiform appendix *(p. 609)*

ascending colon *(p. 609)*

transverse colon *(p. 609)*

descending colon *(p. 609)*

sigmoid colon *(p. 609)*

taeniae coli *(p. 609)*

DENTAL ASSISTANT

Dental assistants prepare patients for dental examination, assist in the examination by handing the dentist the proper instruments and materials, operate suction and other devices to keep the patient's mouth clear during examination, and process x-ray film. They are trained to prepare for and perform some dental procedures, such as molded impressions of teeth, application of medication, and preparing and cleaning teeth. Dental assistants also give instructions on oral health, and many assistants manage the office and do all types of clerical work. The dental assistant differs from the dental hygienist—a trained professional who has received a license to scale and polish teeth.

Most assistants work in private offices, but others work in dental schools, public health departments, and private clinics. Many assistants have on-the-job training, but an increasing number of dental assistants go through one- to two-year post–high-school programs. A certificate, although helpful, is not required for employment.

Based on the terminology contained within this chapter, label the following figures: **LABELING ACTIVITY**

1   Organs and parts of the digestive system.

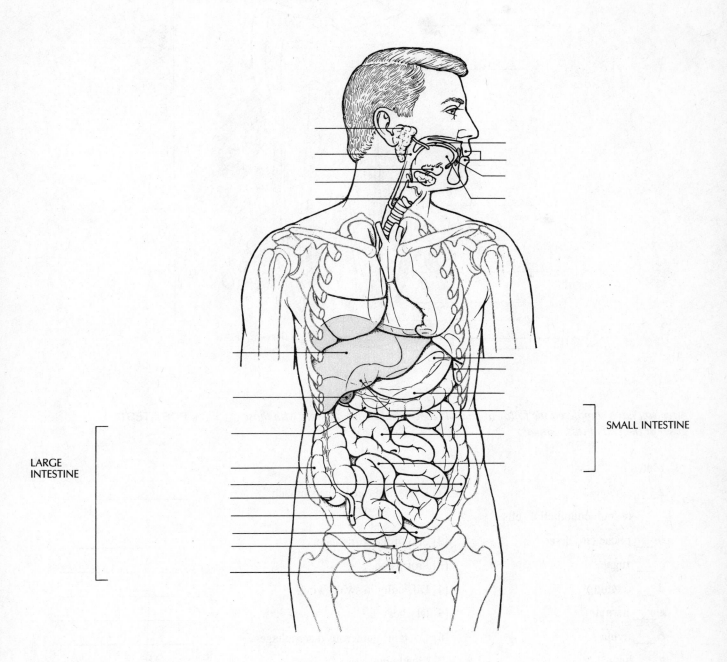

SMALL INTESTINE

LARGE
INTESTINE

2   Sagittal section of a tooth.

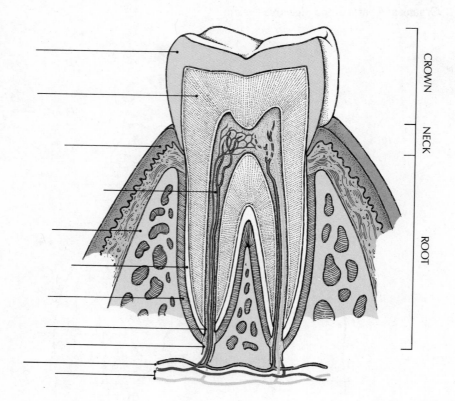

After you have completed the following activities, compare your answers with those given at the end of this manual.

POST-TEST

1   Matching.

a ___ stellate reticuloendothelial cells   [1] Folds of tissue in the stomach.

b ___ plicae circulares   [2] Part of the soft palate.

c ___ rugae   [3] Protrusion.

d ___ retching   [4] Difficulty in swallowing.

e ___ hernia   [5] Dry heaves.

f ___ uvula   [6] Contains numerous macrophages.

g ___ hiatus   [7] Lined with villi.

h ___ dysphagia   [8] Protective and insulating membrane.

i ___ omentum   [9] Gap.

j ___ lamina propria   [10] Macrophages.

2 All of the following are anatomical sphincters except the: _____

  a external anal sphincter.

  b internal anal sphincter.

  c pyloric sphincter.

  d rugal sphincter.

  e a and d.

3 The vagus nerve: _____

  a inhibits stomach motility.

  b begins the cephalic phase of gastric secretion.

  c reduces peristaltic activities.

  d a and b.

  e a, b, and c.

4 The _____ is a ring of smooth muscle and connective tissue between _____

  the terminal antrum of the stomach and the duodenum.

5 Hydrochloric acid is secreted by the _____ cells of the stomach. _____

6 Pepsin is secreted in an inactive form known as _____ . _____

7 The _____ stored in the _____ emulsifies fats. _____

  _____

8 True or False.

  i Teeth are the hardest and most chemically stabilized tissues in the body. _____

  a true

  b false

  ii The entire esophagus is surrounded by smooth muscle. _____

  a true

  b false

  iii One function of saliva is to moisten and lubricate food. _____

  a true

  b false

  iv In the absence of bile, feces are a yellowish brown. _____

  a true

  b false

  v The most important factor controlling gastric emptying is the gastric volume. _____

  a true

  b false

9 Which of the following is *not* a major GI hormone?

   a gastrin

   b epinephrine

   c secretin

   d CCK

   e enterogasterone

10 A double fold of peritoneum suspending the stomach from the liver is called the

                .

11 Bands of longitudinal muscle along the colon called   create a

    string of pouched sacs called   .

12 The gallbladder is primarily responsible for:

   a manufacturing bile.

   b storing bile.

   c diluting bile.

   d concentrating bile.

   e both **b** and **d.**

13 The carpet of fingerlike projections on the inner surface of the duodenum is called:

   a rugae.

   b plicae circulares.

   c villi.

   d crypts.

   e none of the above.

14 One structure that the prancreas is *not* normally in contact with is the:

   a left kidney.

   b duodenum.

   c stomach.

   d inferior vena cava.

   e aorta.

15 Which of the following structures is present in the portal area of the liver?

   a hepatic vein

   b hepatic artery

   c hepatic portal vein

   d bile ducts

   e all of the above

16 Which of the following is *not* an enzyme secreted by the stomach cells?   _____

   a gastrin

   b pepsinogen

   c secretin

   d lipase

   e motilin

17 True or False.

   i Gallstones are technically the same as kidney stones.   _____

   a true

   b false

   ii Hepatitis means liver inflammation.   _____

   a true

   b false

   iii The hormone that stimulates the release of gastric secretions is called secretin.   _____

   a true

   b false

   iv Chief cells produce HCl.   _____

   a true

   b false

   v Jaundice can result from the accumulation of bilirubin in the blood.   _____

   a true

   b false

18 The nasopharyngeal passageway is blocked by the _____ and soft   _____

   palate during swallowing.

19 Salivary amylase catalyzes the breakdown of polysaccharides to: _____

   a disaccharides.

   b carbohydrates.

   c monosaccharides.

   d a and b.

   e a, b, and c.

20 A set of permanent teeth consists of: _____

   a 18 teeth.

   b 20 teeth.

   c 32 teeth.

   d 36 teeth.

   e 42 teeth.

21 The teeth that grind food are called _____ .

                                                                        _____

22 The _____ are wrinkles in the inner mucosal layer of the stomach.

                                                                         _____

23 The salivary gland that secretes a watery serous fluid and an enzyme is the:

    a submaxillary.

    b sublingual.

    c submandibular.

    d parotid.

24 The extrinsic control of gastric function has three phases known as the cephalic, gastric, and absorptive phases.

                                                                          _____

    a true

    b false

25 The cells that produce bile are called _____ .

                                                                          _____

The Urinary System

24

Prefixes

calyc-	a small cup
corpusc-	little body
cort-	covering
fenestr-	opening; window
glomer-	in the form of a ball
juxta-	beside; near
lamin-	layer
mictur-	urinate
nephro-	kidney
papill-	nipple
pod-	foot
rect-	straight
ren-	kidney
trigon-	triangular shape
ure-	urine
vasa-	vessel

Suffixes

-cyte	hollow vessel
-itis	inflammation
-lobar	a lobe
-uria	condition of urine

Read Chapter 24 in the textbook, focusing on the major objectives. As you progress through it with your textbook open, answer and/or complete the following. When you complete this exercise, you will have a thorough outline of the chapter.

DEVELOPING YOUR OUTLINE

I. Introduction (p. 622)

1 What are some functions of the urinary system?

II. The Urinary System: Components and Functions (p. 622)

OBJECTIVE 1

1 Describe the four major components of the urinary system and list a specific function of each.

a

b

c

d

2 What three physiological functions are accomplished by the nephron?

   a

   b

   c

OBJECTIVE 2

OBJECTIVE 3

III. Anatomy of the Kidneys (p. 622)

  A Location and External Anatomy (p. 622)

    1 Where are the kidneys located?

    2 From the inside to the outside, describe the three layers of the kidney.

     a

     b

     c

OBJECTIVE 4

  B Internal Anatomy (p. 624)

    1 From the inside out, describe the three regions of the kidneys.

     a

     b

     c

    2 What are the two regions of the renal cortex?

     a

     b

    3 Describe the composition of the renal columns.

OBJECTIVE 5

  C Blood Supply (p. 626)

    1 Describe the pathway blood takes as it passes through the various vessels of the kidney.

2 About what percent of the cardiac output passes through the kidneys each minute?

3 What is the physiological role of the vasa recta?

D Lymphatics (p. 627)
   1 What is the general course of lymphatic vessels within the kidney?

E Nerve Supply (p. 627)
   1 How do the renal nerves help control kidney function?

F The Nephron (p. 627)   **OBJECTIVE 6**
   1 What are the four different parts of the tubular component of the nephron? Describe a function of each component.

      a

      b

      c

      d

   2 What three physiological processes occur within the nephron?
      a

      b

      c

   3 Contrast locations of the cortical and juxtamedullary nephrons.

G Glomerular (Bowman's) Capsule (p. 627)
   1 Describe the structures that form the renal corpuscle.

   2 Name the three layers of the endothelial capsular membrane.
      a

b

c

3 Specifically, where does filtration take place within the renal corpuscle?

4 Describe the function of the proximal convoluted tubule.

5 Which part of the loop of the nephron (loop of Henle) is permeable to water?

6 The distal convoluted tubule actively transports _____ and _____ ions into the filtrate.

7 How do the collecting ducts concentrate the glomerular filtrate?

8 List and describe the four different anatomical parts of the juxtaglomerular apparatus.

| OBJECTIVE 7 |

a

b

c

d

IV. Accessory Excretory Structures (p. 632)

| OBJECTIVE 8 |

A Ureters (p. 632)

1 Describe the function of the ureters.

2 The three layers of the ureter, from the inside out, are the:

a

b

c

B Urinary Bladder (p. 632)

   1 Describe the function of the urinary bladder.

   2 How does the histology of the urinary bladder differ from the ureters?

   3 Name the sphincter muscles associated with the bladder.

C Urethra (p. 634)

   1 Name the three portions of the male urethra. Briefly describe the significance of each portion.

     a

     b

     c

   2 How does the urethra in the male differ from the urethra in the female?

V. The Effects of Aging on the Urinary System (p. 634)

OBJECTIVE 9

   1 What are some problems that occur with the urinary system with age?

VI. Developmental Anatomy of the Kidneys (p. 634)

   1 Describe how the kidneys develop.

VII. When Things Go Wrong (p. 636)

   1 Briefly describe each of the following disorders:

     a acute renal failure

     b chronic renal failure

c glomerulonephritis

d pyelonephritis

e renal calculi

f cystitis

g urethritis

h urinary incontinence

i kidney cancer

j bladder cancer

k nephroptosis

These are the terms you should know before proceeding to the post-test.   **MAJOR TERMS**

excretion *(p. 622)*	interlobar arteries *(p. 626)*
urinary system *(p. 622)*	arcuate arteries *(p. 626)*
kidneys *(p. 622)*	interlobular arteries *(p. 627)*
hilus *(p. 622)*	afferent arteries *(p. 626)*
renal pelvis *(p. 624)*	glomerulus *(p. 627)*
major calyces *(p. 624)*	efferent arterioles *(p. 627)*
minor calyces *(p. 624)*	peritubular capillaries *(p. 627)*
renal medulla *(p. 624)*	interlobular veins *(p. 627)*
renal pyramids *(p. 624)*	arcuate veins *(p. 627)*
renal cortex *(p. 624)*	interlobar veins *(p. 627)*
renal columns *(p. 626)*	renal vein *(p. 627)*
kidney tubules *(p. 626)*	arteriae rectae *(p. 627)*
renal artery *(p. 626)*	venae rectae *(p. 627)*

vasa recta *(p. 627)*

renal plexus *(p. 627)*

nephron *(p. 627)*

glomerular (Bowman's) capsule *(p. 627)*

renal corpuscle *(p. 627)*

podocytes *(p. 630)*

pedicels *(p. 630)*

filtration slit *(p. 630)*

endothelial capsular membranes *(p. 630)*

glomerular filtrate *(p. 630*

proximal convoluted tubule *(p. 630)*

loop of the nephron (loop of Henle) *(p. 630)*

distal convoluted tubule *(p. 631)*

collecting duct *(p. 631)*

juxtaglomerular cells *(p. 631)*

macula densa *(p. 631)*

juxtaglomerular apparatus *(p. 632)*

ureter *(p. 632)*

tunica mucosa *(p. 632)*

tunica muscularis *(p. 632)*

tunica adventitia *(p. 632)*

urinary bladder *(p. 632)*

tunica serosa *(p. 632)*

trigone (p. *632)*

urethra *(p. 634)*

prostatic portion *(p. 634)*

membranous portion *(p. 634)*

spongy portion *(p. 634)*

metanephros *(p. 634)*

acute renal failure *(p. 636)*

chronic renal failure *(p. 636)*

acute gomerulonephritis *(p. 636)*

chronic glomerulonephritis *(p. 636)*

pyelonephritis *(p. 636)*

renal calculi *(p. 636)*

cystitis *(p. 637)*

urethritis *(p. 637)*

urinary incontinence *(p. 637)*

kidney cancer *(p. 637)*

bladder cancer *(p. 637)*

nephroptosis *(p. 637)*

Based on the terminology contained within the chapter, label the following figures: **LABELING ACTIVITY**

1   A kidney.

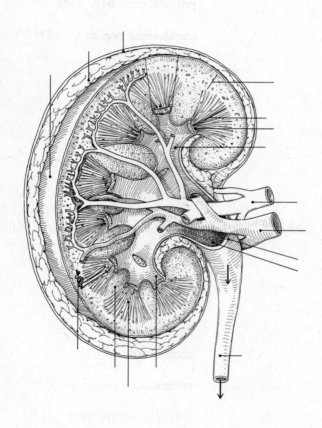

2   The bladder and urethra in a female.

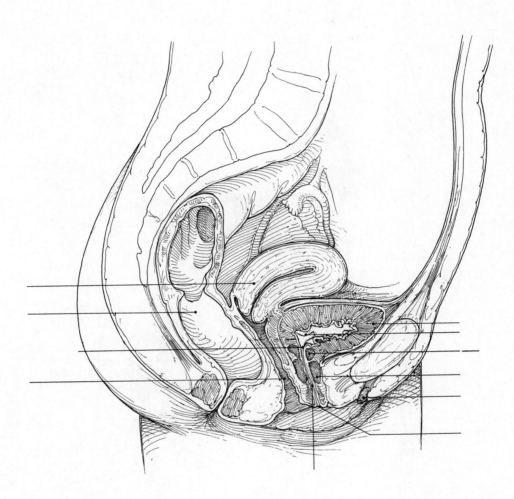

3   The bladder and urethra in a male.

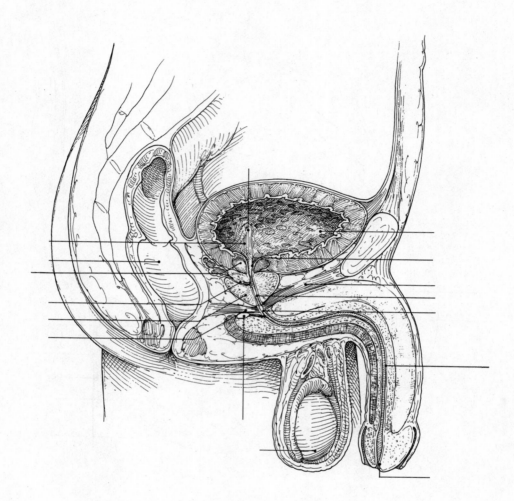

4   Trigone and urethral sphincters.

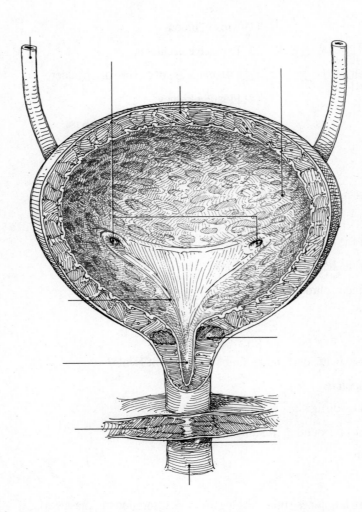

After you have completed the following activities, compare your answers with those given at the end of this manual.

1 Matching.

   a ___ dysuria   [1] Urea in blood.   _____

   b ___ cystitis   [2] Excessive urination.   _____

   c ___ glomerulonephritis   [3] Inflammation of the urinary bladder.   _____

   d ___ polyuria   [4] Floating kidney.   _____

   e ___ nephroptosis   [5] Inflammation of the renal pelvis.   _____

   f ___ pyelitis   [6] Inflammation of the glomeruli.   _____

   g ___ uremia   [7] Painful urination.   _____

2 In which portion of the kidney does salt pass into the surrounding tissue by diffusion?   _____

   a descending loop of the nephron (loop of Henle)

   b bend of the loop of the nephron (loop of Henle)

   c ascending loop of the nephron (loop of Henle)

   d all of the above

   e none of the above

3 Which event takes place in the descending limb of the loop of Henle?   _____

   a active removal of sodium

   b removal of sodium by diffusion

   c passage of water into surrounding fluid by osmosis

   d concentration of urea

   e **c** and **d**

4 Which of the following hormones stimulates the active reabsorption of potassium ions within the kidneys?   _____

   a ADH

   b ACTH

   c aldosterone

   d calcitonin

   e renin

5 What are five major substances regulated by the kidney?   _____

   a

   b

   c

   d

   e

6 The amount of fluid filtered from the plasma into glomerular (Bowman's) capsule per day is approximately _____ liters.

7 The average person excretes between _____ and _____ of urine per day.

8 Which of the following waste materials is/are removed by the kidneys?

   a urea

   b excess salts

   c uric acid

   d a and b

   e a, b, and c

9 Which of the following waste materials is formed by the ammonia that results from protein metabolism?

   a urea

   b bicarbonates

   c uric acid

   d a and b

   e a, b, and c

10 The _____ transfers urine from the kidneys to the bladder.

11 Which bladder sphincter(s) is/are under conscious control?

   a internal

   b external

   c detrusor

   d trigone

   e both c and d

12 The renal corpuscle is the glomerulus and _____ .

13 Which capillary network surrounds the loop of the nephron?

   a glomerulus

   b vasa recta

   c peritubular capillaries

   d a and b

   e a, b, and c

14 Which of the following is involved in urine formation by the human kidney?

   a tubular secretion

   b tubular reabsorption

   c glomerular filtration

   d urea synthesis

   e all but d

15 Renal calculi:

    a are made up of salts of calcium, uric acid, and cysteine.

    b are called kidney stones.

    c contain urea salts.

    d a and b.

    e a, b, and c.

16 Select the more acid of the two items.

    a normal urine pH

    b normal blood pH

The Reproductive Systems

25

Prefixes

andro-	male
ejaculo-	to throw out
follic-	small bag
gyn-	female
haplo-	single
labi-	lip
lacto-	milk
liga-	bind; to tie
mast-	breast
mens-	month
menstru-	monthly
metr-	uterus
myo-	muscle
ov-	egg
puber-	adult
semen-	seeds
semin-	semen
super-	over; above

Suffixes

-arche	beginning; origin
-cide	to kill
-fer	to bear
-gen	origin

Read Chapter 25 in the textbook, focusing on the major objectives. As you progress through it, with your textbook open, answer and/or complete the following. When you complete this exercise, you will have a thorough outline of the chapter.

DEVELOPING YOUR OUTLINE

I. Introduction (p. 642)

1 What is the major benefit of sexual reproduction?

II. Male Reproductive Anatomy (p. 642)

OBJECTIVE 1

1 Describe the four major structures associated with the male reproductive system.

a

b

c

d

A Testes (p. 642)

   1 When do the testes descend into the scrotum?

   2 What might result if the inguinal canal fails to close properly?

   3 Why do the testes hang outside of the body?

   4 Describe the three tunica of the testes.

      a

      b

      c

   5 Complete the following table on the different parts of the seminiferous tubules:

Structure	Function
a _____	Develop into mature sperm cells.
sustentacular cells	**b** _____
c _____	Secrete androgens.

B Sperm (Spermatozoa) (p. 645)

   1 Describe the three anatomical areas of a sperm.

      a

      b

      c

   2 What is the function of the acrosome?

C Accessory Ducts (p. 646)

   1 List the ducts that sperm travel through, beginning with the seminiferous tubules and ending with the seminal vesicle.

      a

b

c

d

e

f

g

h

2 Describe the three main functions of the epididymis.

a

b

c

3 Describe the three anatomical areas of the epididymis.

a

b

c

4 What is the main function of the ductus deferens?

5 What do the ejaculatory ducts eventually form?

6 List and describe the three portions of the male urethra.

a

b

c

7 What is the function of the urethral glands?

8 How are male accessory ducts related?

D Accessory Glands (p. 647)

   1 Complete the following table on the accessory glands of the male:

Accessory gland	Location	Function
a _____	Lie next to the ductus deferens.	Provide bulk of seminal fluid.
b _____	Lies in front of the urinary bladder.	**c** _____
bulbourethral	**d** _____	Fluid acts as a lubricant.

   2 List and describe the three types of glands inside the prostate.

     a

     b

     c

   3 What glands secrete prostaglandins?

E Semen (p. 648)

   1 List the components of semen.

   2 The average amount of semen per ejaculation is about _____ .

   3 The average number of sperm per ejaculation is approximately

   _____ .

F Penis (p. 649)

  1 Describe the two major functions of the penis.

    a

    b

  2 What are the cylindrical strands of erectile tissue within the penis called?

    _____

  3 Fill in the blanks in the following paragraph on the penis.

    The sensitive region of the penis is the **a** _____ . The proximal and very sensitive edge of the glans penis is the **b** _____ . The loosely fitting foreskin over the glans is the **c** _____ . This is removed in a procedure called a **d** _____ . The **e** _____ glands form a cheeselike substance called **f** _____ . The two parts of the CNS that control an erection are the **g** _____ and sacral plexus. When an erect penis returns to its flaccid state, this is termed **h** _____ .

III. Female Reproductive Anatomy (p. 650)

  1 Name the organs of the female reproductive system and give a function of each.

    a

    b

    c

    d

    e

    f

A Ovaries (p. 650)

   1 Describe the ovaries according to:

    a size and shape _____

    b location _____

    c attachment _____

    d internal structure _____

   2 How are ovarian follicles classified?

   3 What is the function of the corpus luteum?

B Uterine tubes (p. 652)

   1 What are two other anatomical names for the uterine tubes?

    a

    b

   2 List and describe the three anatomical portions of the uterine tubes.

    a

    b

    c

   3 List and describe the three layers of the uterine tube wall.

    a

    b

    c

   4 What structural features of the uterine tubes enhance passage of the ova both into and out of the tube?

   5 What is an ectopic pregnancy?

C Uterus (p. 653)

   1 Complete the following table on the uterine ligaments:

Ligament(s)	Function
a _____	Keep the uterus tilted over the bladder.
b _____	Attached to the lateral wall of the pelvis.
uterosacral	**c** _____ _____
posterior	**d** _____ _____
e _____	Attaches uterus to urinary bladder.

   2 In the space below, draw and label the uterus. Label: *fundus, cervix, cervical canal, uterine cavity, serous coat, myometrium,* and *endometrium.* Briefly describe the functional significance of each feature.

D Vagina (p. 654)

   1 Describe three functions of the vagina.

     a

     b

     c

2 Describe the following vaginal structures:

   a rugae

   b hymen

E External Genital Organs (p. 655)
   1 Describe the following structures:

   a mons pubis

   b labia majora

   c labia minora

   d clitoris

   e vestibule

   f greater vestibular glands

   g lesser vestibular glands

   h perineum

F Mammary Glands (p. 656)   **OBJECTIVE 8**
   1 What determines the size of the breasts?

   2 What ligaments hold the breasts in place?

   3 Describe the duct system of the breast.

IV. Formation of Sex Cells (Gametogenesis) (p. 656)

OBJECTIVE 9

A Spermatogenesis (p. 656)

    1 Fill in the blanks in the following paragraph on sperm formation:

At the time of puberty, **a** _____ become activated by testosterone. Each spermatogonium divides by **b** _____ to produce two daughter cells: a primary spermatogonium and a

c _____ . The primary spermatocyte undergoes

d _____ and produces two smaller **e** _____ . Both of these cells then undergo a second **f** _____ division to form the **g** _____ , which develops into mature **h** _____

B Oogenesis (p. 657)

    1 Fill in the blanks in the following paragraph of ovum formation:

The basic cell of the ovum is the **a** _____ , which develops into a primary **b** _____ , which in turn undergoes meiosis to produce a large **c** _____ and a small **d** _____ . If the large secondary oocyte is fertilized, it undergoes a second

e _____ division, is reduced to the haploid number of chromosomes, and is called an **f** _____ .

    2 How does oogenesis differ from spermatogenesis?

V. Conception (p. 658)

OBJECTIVE 10

A Fertilization (p. 658)

    1 Describe fertilization.

    2 Name the usual site of fertilization.

    3 Fertilization must occur _____ hours after ovulation.

    4 Sperm can remain viable for about _____ hours.

    5 How many sperm fertilize an ovum? _____

    6 What structure of the sperm releases hyaluronidase?

    7 What is the function of hyaluronidase?

    8 What prevents more than one sperm from entering an ovum?

9 If more than one sperm penetrates an ovum, this is termed

_____

_____ .

B Human Sex Determination (p. 659)

    1 How does the male determine the sex of a child?

    2 When is the sex of a child determined?

VI. The Effects of Aging on the Reproductive Systems (p. 659)

    1 As a female ages, what happens to her reproductive system?

    2 As a male ages, what happens to his reproductive system?

VII. Developmental Anatomy of the Reproductive Systems (p. 662)

    1 How do the reproductive systems develop?

VIII. When Things Go Wrong (p. 664)

OBJECTIVE 11

    1 What agent causes the following STDs?

        a nongonococcal urethritis

        b type II herpes simplex

        c gonorrhea

        d syphilis

        e trichomoniasis

        f pelvic inflammatory disease

    2 Describe the following diseases:

        a ovarian cysts

        b follicular cysts

        c granulosa-lutein cysts

        d endometriosis

These are terms you should know before proceeding to the post-test.

MAJOR TERMS

testes *(p. 642)*

scrotum *(p. 642)*

tunica albuginea *(p. 642)*

seminiferous tubules *(p. 642)*

spermatogenic cells *(p. 642)*

spermatogenesis *(p. 642)*

sustentacular cells *(p. 642)*

interstitial endocrinocytes *(p. 645)*

androgens *(p. 645)*

testosterone *(p. 645)*

sperm (spermatozoa) *(p. 645)*

sterile *(p. 646)*

tubuli recti *(p. 646)*

rete testis *(p. 646*

efferent ducts *(p. 646)*

epididymis *(p. 646)*

ductus deferens *(p. 646)*

ejaculatory duct *(p. 646)*

urethra *(p. 646)*

prostatic urethra *(p. 647)*

membranous urethra *(p. 647)*

spongy urethra *(p. 647)*

seminal vesicles *(p. 647)*

prostate gland *(p. 648)*

bulbourethral glands *(p. 648)*

semen *(p. 648)*

penis *(p. 649)*

corpora cavernosa *(p. 649)*

corpus spongiosum *(p. 649)*

glans penis *(p. 649)*

prepuce (foreskin) *(p. 649)*

circumcision *(p. 649)*

ovaries *(p. 650)*

mesovarium *(p. 650)*

broad ligament *(p. 650)*

ovarian ligament *(p. 650)*

hilum *(p. 650)*

suspensory ligament *(p. 650)*

germinal layer *(p. 650)*

stroma *(p. 650)*

follicles *(p. 650)*

tunica albuginea *(p. 650)*

primordial follicles *(p. 650)*

vesicular ovarian follicles *(p. 650)*

corpus luteum *(p. 650)*

uterine tubes *(p. 652)*

infundibulum *(p. 652)*

ampulla *(p. 652)*

isthmus *(p. 652)*

fimbriae *(p. 653)*

uterus *(p. 653)*

round ligaments *(p. 653)*

broad ligaments *(p. 653)*

uterosacral ligaments *(p. 653)*

posterior ligament *(p. 653)*

anterior ligament *(p. 653)*

fundus *(p. 653)*

cervix *(p. 653)*

body of uterus *(p. 653)*

isthmus *(p. 653)*

cervical canal *(p. 653)*

uterine cavity *(p. 653)*

myometrium *(p. 653)*

endometrium *(p. 653)*

menstrual flow *(p. 654)*

menstruation *(p. 654)*

vagina *(p. 654)*

rugae *(p. 654)*

external genitalia *(p. 655)*

vulva *(p. 655)*

mons pubis *(p. 655)*

labia majora *(p. 655)*

labia minora *(p. 655)*

clitoris *(p. 655)*

glans *(p. 655)*

vestibule *(p. 655)*

greater vestibular glands *(p. 655)*

lesser vestibular glands *(p. 655)*

perineum *(p. 655)*

mammary glands *(p. 656)*

prolactin *(p. 656)*

lactiferous ducts *(p. 656)*

lactiferous sinus *(p. 656)*

spermatogenesis *(p. 656)*

spermatogonia *(p. 656)*

primary spermatocyte *(p. 657)*

secondary spermatocytes *(p. 657)*

spermatids *(p. 657)*

oogenesis *(p. 657)*

oogonium *(p. 658)*

primary oocyte *(p. 658)*

secondary oocyte *(p. 658)*

first polar body *(p. 658)*

second polar body *(p. 658)*

ootid *(p. 658)*

mature ovum *(p. 658)*

zygote *(p. 658)*

conception *(p. 658)*

acrosin *(p. 658)*

zona pellucida *(p. 658)*

vitelline membrane *(p. 658)*

fertilization membrane *(p. 658)*

autosomes *(p. 659)*

sex chromosomes *(p. 659)*

sexually transmitted diseases (STD) *(p. 664)*

nongonococcal urethritis *(p. 664)*

type II herpes simplex *(p. 664)*

gonorrhea *(p. 664)*

syphilis *(p. 664)*

trichomoniasis *(p. 665)*

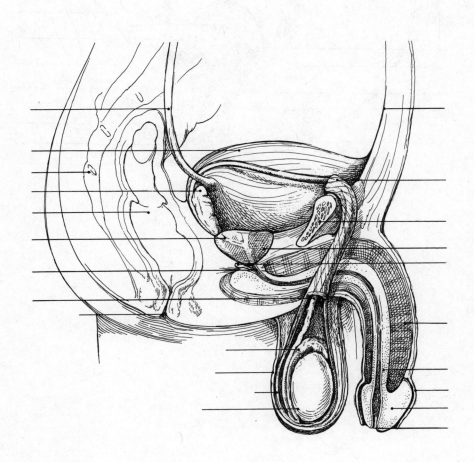

pelvic inflammatory disease (PID) *(p. 665)*

breast cancer *(p. 665)*

cervical cancer *(p. 665)*

prostatic cancer *(p. 665)*

benign prostatic hypertrophy (BPH) *(p. 665)*

ovarian cysts *(p. 666)*

endometriosis *(p. 666)*

Using the terminology contained within the chapter, label the following figures: **LABELING ACTIVITY**

1   The male reproductive system.

ACCESSORY DUCTS

ACCESSORY GLANDS

PENIS

2   The female reproductive system.

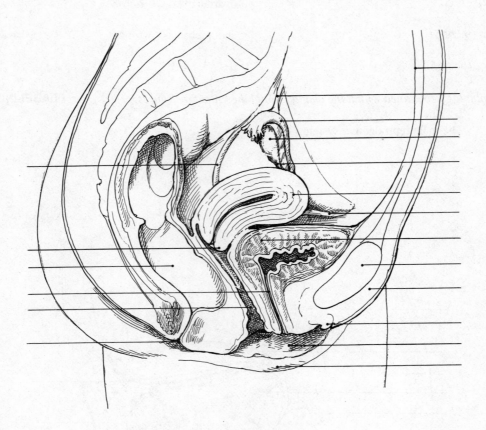

3   The ovary in section.

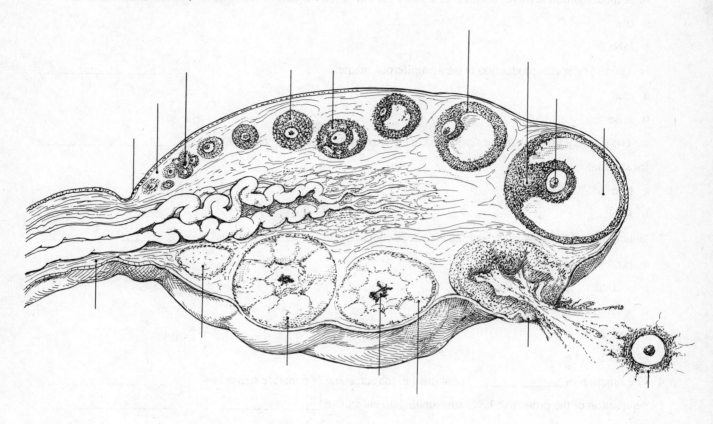

After you have completed the following activities, compare your answers with those given at the end of this manual

1 a During spermatogenesis, meiosis takes place between the _____ and

   the _____ .

   b Testosterone is produced by interstitial cells that are located between the

   _____ .

   c Ejaculation is usually accompanied by an intense pleasure known as an

   _____ .

2 True or False.

   i Implantation of the blastocyst occurs 24 hours after fertilization.

   a true

   b false

   ii Once a sperm penetrates an ovum, the zona pellucida's chemical constitution is

   altered so that another sperm cannot penetrate.

   a true

   b false

iii A tubal ligation involves removal of a small section of the vas deferens.   _____

   a true

   b false

iv The site of sperm production is the seminiferous tubules.   _____

   a true

   b false

v The primary function of the scrotum is to ensure constant testicular temperature   _____

for spermatogenesis.

   a true

   b false

3 Which of the following male reproductive cells contains a single haploid nucleus?   _____

   a primary spermatocytes

   b secondary spermatocytes

   c spermatids

   d both **a** and **b**

   e both **b** and **c**

4 The function of _____ contained in the acrosome of a mature sperm is   _____

penetration of the protective layer surrounding an unfertilized _____ .   _____

5 Most of the sperm produced by the testes are stored in the: _____

   a seminal vesicles.

   b bulbourethral glands.

   c Cowper's glands.

   d vas deferentia and epididymis.

   e seminiferous tubules.

6 Which of the following tissues contributes fluid(s) to semen?   _____

   a epididymis

   b bulbourethral glands

   c seminal vesicles

   d prostate gland

   e all of the above

7 Meiosis I is not complete in the female until: _____

   a menopause.

   b birth.

   c fertilization.

   d menstruation.

   e ovulation.

8 At the completion of meiosis I, _____ are present within an ovarian follicle.

   a a secondary oocyte and a polar body

   b a primary oocyte and a polar body

   c an ovum and three polar bodies

   d two oocytes and no polar body

   e oogonia but no polar bodies

9 Which of the following is (are) *not* part of the vulva?

   a labia minora

   b labia majora

   c clitoris

   d urethra

   e both **a** and **b**

10 Sperm gain motility while they are in the:

   a urethra.

   b vas deferens.

   c epididymis.

   d seminiferous tubules.

   e bulbourethral glands.

11 Genital herpes:

   a can be treated with penicillin.

   b produces blisters on the genitals.

   c can be transmitted to a baby during childbirth.

   d both **b** and **c.**

12 The male fetus is genetically represented as:

   a XX.

   b XY.

   c YY.

   d ZZ

   e ZX.

13 Matching.

   a ___ testes   [1] Outside of the seminiferous tubules.

   b ___ scrotum   [2] The male gonads.

   c ___penis   [3] Spermatogenesis occurs.

   d ___ seminiferous tubules   [4] Deposits sperm.

   e ___interstitial endocrinocytes   [5] Contains the testes and spermatic

                                        cord.

14 Matching.

   a ___ testosterone

   b ___ efferent duct

   c ___ epididymis

   d ___ ejaculation

   e ___ vas deferens

[1] Coiled ducts. _____

[2] A male hormone. _____

[3] Sudden emission of semen. _____

[4] An extension of the epididymus. _____

[5] Carries sperm to the epididymus. _____

15 Matching.

   a ___ inguinal canal

   b ___ ejaculatory duct

   c ___ seminal vesicle

   d ___ prostate gland

   e ___ vas deferens

[1] Two pea-size bulbourethral glands. _____

[2] Surrounds the neck of the bladder. _____

[3] Secretes 60 percent of the seminal fluid. _____

[4] Passage through which semen enters theurethra. _____

[5] A passageway to the scrotum. _____

16 Matching.

   a ___ ovaries

   b ___ uterine tubes

   c ___ uterus

   d ___ vagina

[1] A muscular tube. _____

[2] The gamete-producing organs of the female. _____

[3] A pear-shaped organ. _____

[4] Carries an ovum to the uterus. _____

17 Matching.

   a ___ mons pubis

   b ___ labia minora

   c ___ clitoris

   d ___ greater vestibular glands

[1] A pad of adipose tissue. _____

[2] Two elongated folds of skin. _____

[3] An erectile structure. _____

[4] Secrete lubricant material. _____

18 Sexual intercourse is often referred to as:

   a copulation. _____

   b coitus.

   c ejaculation.

   d orgasm.

   e both **a** and **b**.

19 Fertilization occurs in the upper one-third of the uterine tube within 24 hours after _____ ovulation.

   a true

   b false

20 Whereas spermatogenesis yields four spermatozoa after meiosis, oogenesis yields two ova.

   a true

   b false

21 With respect to STDs, the most common sexually transmitted organism is:

   a *Neisseria gonorrhoreae.*

   b *Chlamydia trachomatis.*

   c *Treponema pallidum.*

   d *Trichomonas vaginalis.*

   e *Ureaplasma urealyticum.*

Developmental Anatomy:
A Life Span Approach
26

Prefixes

allant-	sausage-shaped
amnio-	amnion
blast-	a bud
chorio-	membrane; skin
cleav-	to divide
gesta-	to bear; carry
miss-	let go; send forth
morul-	mulberry
nata-	birth
neo-	new
orgasm-	to swell
post-	after
troph-	nourishment
ultra-	beyond
umbil-	navel

Suffixes

-cele	hollow
-centesis	a piercing
-chord	a string
-cyst	bladder
-derm	skin
-genic	genes
-scope	observing

Read Chapter 26 in the textbook, focusing on the major objectives. As you progress through it, with your textbook open, answer and/or complete the following. When you complete this exercise, you will have a thorough outline of the chapter.

DEVELOPING YOUR OUTLINE

I. Introduction (p. 671)

OBJECTIVE 1

  1 Fill in the blanks in the following paragraph on prenatal development:

    During the prenatal period, a developing individual is called a

    _____ for the first week of existence, and _____

    for the next seven weeks, and a _____ from nine weeks until

    birth.

  2 When does a developing individual become an infant?

  3 Describe the two periods of prenatal development.

    a

    b

II. Embryonic Development (p. 671)

    1 What resumes in the ovum after entry of a sperm?

A Cleavage (p. 671)

OBJECTIVE 2

    1 What forms immediately after a sperm penetrates an ovum?

    _____

    2 After the first cleavage of the zygote, what are the cells called?

    _____

    3 Next, the cells form a structure called a(n) _____ .

    4 Fill in the blanks in the following paragraph on cleavage:

    A morula will eventually develop a _____ with a covering

    mass of cells called a _____ , which eventually forms the

    _____ . The embryo actually grows from an inner cell mass

    called a(n) _____ .

B Implantation in the Uterus (p. 671)

OBJECTIVE 3

    1 When does pregnancy actually begin?

    2 Discuss the various chemicals that make implantation possible.

    3 What two layers comprise the bilaminar embryonic disk?

    a

    b

C Extraembryonic Membranes (p. 673)

OBJECTIVE 4

    1 Where did the extraembryonic membranes get their name?

    2 Describe the function of each of the following:

    a chorionic villi

    b chorion

    c yolk sac

    d yolk stalk

e amnion

f amniotic fluid

g allantois

D Placenta and Umbilical Cord (p. 675)   OBJECTIVE 5
  1 Describe the three main functions of the placenta.
  a

  b

  c

  2 Describe the structure and functions of the umbilical cord.

E Weeks 1 Through 8: Formation of Body Systems (p. 678)   OBJECTIVE 6
  1 Complete the following table on the major developments that occur
    during the embryonic period:

Time period	Major developments
week 1	_____
week 2	_____
week 3	_____
week 4	_____
week 5	_____
week 6	_____
week 7	_____
week 8	_____

III. Fetal Development (p. 680)

1 Write the number of the lunar month (embryonic or fetal period) when each of the following events occurs:

a _____ The main blood vessels assume their final organization.

b _____ Full term occurs.

c _____ Nails reach the tips of the fingers and toes.

d _____ External genitalia attain their distinctive features.

e _____ The face looks human.

f _____ The testes settle into the scrotum.

g _____ Lanugo hair is very prominent.

h _____ Thumb sucking may begin.

2 Describe the fetal development of the major body structures.

IV. Maternal Events of Pregnancy (p. 683)

1 What is usually the first sign of pregnancy?

2 What is Hagar's sign?

3 Write the number of week(s) when each of the following events occurs:

a _____ The fetal heart can be heard with a stethoscope.

b _____ Normal birth occurs.

c _____ Mother feels the fetus moving.

d _____ Menstruation ceases.

e _____ Breasts begin to enlarge.

f _____ Morning sickness begins.

g _____ The abdomen begins to protrude.

A Pregnancy Testing (p. 683)

1 What hormone is the basis for most pregnancy tests? _____

V. Birth and Lactation (p. 670)

1 Parturition usually occurs about _____ days after fertilization or about _____ days from the first day of the menstrual period preceding fertilization.

A Process of Childbirth (p. 685)

    1 Initial uterine contractions are stimulated by the hormone _____.

    2 Babies are usually born in the _____ position.

    3 Describe the three stages of labor.

        a

        b

        c

B Premature and Late Birth (p. 686)

    1 Describe some characteristics of a premature baby.

    2 What are some problems associated with a premature birth?

C Multiple Births (p. 687)

    1 Describe the development and characteristics of fraternal twins.

    2 Describe the development and characteristics of identical twins.

    3 What are conjoined twins?

    4 What are some causes of multiple births?

D Adjustments to Life Outside the Uterus (p. 688)

    1 What are some of the adjustments that a baby must make at birth to enable life outside the uterus?

OBJECTIVE 10

VI. Postnatal Life Cycle (p. 689)

OBJECTIVE 11

    1 Describe the three processes involved in prenatal development.

        a

        b

        c

A Neonatal Stage (p. 689)

    1 Other than respiratory and circulatory adjustments, what other problems does the neonate face?

B Infancy (p. 690)

    1 How long does infancy last? _____

C Childhood (p. 690)

    1 How long does childhood last? _____

D Adolescence and Puberty (p. 690)

    1 Describe adolescence and puberty.

    2 The first menstrual period is called the _____ .

E Adulthood (p. 691)

    1 Adulthood spans the years between about _____ and

        _____ and old age.

    2 The cessation of menstrual periods is called _____ .

F Senescence (p. 692)

    1 The indeterminate period when an individual is said to grow old is called

        _____ .

VII. When Things Go Wrong (p. 692)

    OBJECTIVE 12, 13

    1 List several disorders that can be detected via amniocentesis.

    2 What is the purpose of fetoscopy?

    3 List several disorders that can be detected via chorionic villi sampling.

    4 How is alpha-fetoprotein used?

    5 What specific disease would warrant an intrauterine transfusion?

    6 Ultrasonography is used to determine _____ .

    7 Describe some reasons for performing fetal surgery.

These are terms you should know before proceeding to the post-test. **MAJOR TERMS**

prenatal *(p. 671)*

postnatal *(p. 671)*

gestation period *(p. 671)*

embryonic period *(p. 671)*

fetal period *(p. 671)*

zygote *(p. 671)*

cleavage *(p. 671)*

blastomeres *(p. 671)*

morula *(p. 671)*

blastocyst *(p. 671)*

blastocoel *(p. 671)*

inner cell mass *(p. 671)*

trophectoderm *(p. 671)*

implantation *(p. 671)*

human chorionic gonadotropin (hCG) *(p. 673)*

bilaminar embryonic disk *(p. 673)*

embryo *(p. 673)*

extraembryonic (fetal) membranes *(p. 673)*

chorionic villi *(p. 675)*

chorion *(p. 675)*

yolk stalk *(p. 675)*

amnion *(p. 675)*

amniotic cavity *(p. 675)*

amniotic fluid *(p. 675)*

allantois *(p. 675)*

placenta *(p. 677)*

umbilical cord *(p. 677)*

afterbirth *(p. 678)*

full term *(p. 678)*

parturition *(p. 685)*

cephalic position *(p. 685)*

labor *(p. 685)*

puerperal period *(p. 686)*

fraternal (dizygotic) twins *(p. 687)*

identical (monozygotic) twins *(p. 688)*

conjoined (Siamese) twins *(p. 688)*

foramen ovale *(p. 689)*

ductus arteriosus *(p. 689)*

growth *(p. 689)*

morphogenesis *(p. 689)*

cellular differentiation *(p. 689)*

development *(p. 689)*

neonate *(p. 689)*

infancy *(p. 690)*

childhood *(p. 690)*

adolescence *(p. 690)*

puberty *(p. 690)*

secondary sex characteristics *(p. 690)*

menarche *(p. 690)*

adulthood *(p. 691)*

menopause *(p. 691)*

senescence *(p. 692)*

amniocentesis *(p. 692)*

fetoscopy *(p. 693)*

chorionic villi sampling *(p. 693)*   ultrasonography *(p. 694)*

alpha-fetoprotein test *(p. 693)*   fetal surgery *(p. 694)*

intrauterine transfusion *(p. 694)*

Using the terminology contained within the chapter, label the following figures: **LABELING ACTIVITY**

1   A blastocyst.

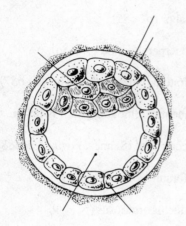

5 TO 6 DAYS AFTER FERTILIZATION
BLASTOCYST (EARLY)

2   Implantation in the uterus.

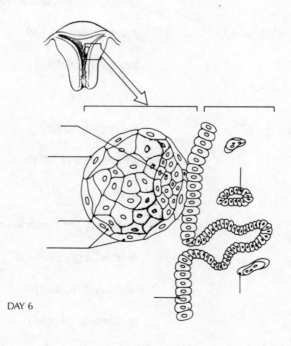

DAY 6

After you have completed the following activities, compare your answers with those given at the end of this manual.

1 Matching.

   a ___ blastocyst   [1] Gives rise to umbilical blood vessels.   _____

   b ___ allantois   [2] First secretion of mammary glands.   _____

   c ___embryonic disk   [3] A hollow ball of cells.   _____

   d ___ chorion   [4] Two germ layers.   _____

   e ___colostrum   [5] Embryonic membrane that contacts   _____

                                               uterine wall.

2 When an embryo is clearly recognizable as a human being it is called a(n)

   _____ .   _____

3 The embryonic membrane that functions to form blood cells is the _____.   _____

4 The first cell of "life" is the _____ .   _____

5 Mesodermal cells give rise to: _____

   a the digestive tract lining.

   b the epidermis.

   c nerve tissue.

   d muscle and bone tissue.

   e both **a** and **b.**

6 Adolescence: _____

   a ends at puberty.

   b occurs at 12 to 16 years of age.

   c is usually over by 16 years of age.

   d extends from before puberty to adulthood.

   e begins at puberty.

7 The neonatal period extends from: _____

   a one to two years of age.

   b birth to four weeks of age.

   c birth to one year of age.

   d conception to birth.

   e birth to puberty.

8 The waste substance found in the fetal colon is called _____ .   _____

9 True or False.

   i   The placental barrier consists of trophoblastic cells of the chorionic villi,   _____

connective tissue, and fetal capillary endothelium.

   a true

   b false

   ii The primary excretory organ throughout fetal life is the kidney.   _____

   a true

   b false

   iii Growth is defined as an increase in body size.   _____

   a true

   b false

   iv Puberty usually begins earlier in males.   _____

   a true

   b false

   v The first menstruation period is termed gestation.   _____

   a true

   b false

10 Which of the following hormones inhibits uterine contractions?   _____

   a oxytocin

   b progesterone

   c estrogen

   d **a** and **b**

   e **a, b,** and **c**

11 Arrange the following terms in correct order from earliest to latest:

morula, fetus, embryo, blastocyst, zygote.

_____ , _____ , _____ , _____ ,

_____ .

12 True or False.

   i   Human ova normally can be simultaneously fertilized by more than one sperm.   _____

   a true

   b false

   ii Cleavage results in an increase in the overall size of the embryo.   _____

   a true

   b false

iii The neurological growth of a developing human is completed by the end of the _____
third trimester of pregnancy.

    a true

    b false

iv The heart is the first organ formed in the human embryo. _____

    a true

    b false

v The placenta develops from the trophoblast. _____

    a true

    b false

13 Which structure does *not* develop from mesoderm? _____

    a muscles

    b gonads

    c lungs

    d blood vessels

    e heart

14 Which event takes place during the second week of human embryonic develop- _____
ment?

    a organogenesis

    b gastrulation

    c cleavage

    d neural crest formation

    e implantation

15 Babies can survive outside of the mother after _____ months of _____
development.

16 The sucking reflex begins in the _____ month of development. _____

17 Lactation occurs: _____

    a as a result of the immediate loss of estrogen after birth.

    b as a result of the immediate loss of progesterone after birth.

    c as a result of prolactin production by the pituitary.

    d a and b.

    e a, b, and c.

18 Which of the following is a characteristic of the third stage of labor?   _____

   a amniotic fluid escapes from the uterus.

   b the placenta is separated from the uterine wall.

   c the placenta is expelled.

   d b and c.

   e a, b, and c.

19 Implantation occurs: _____

   a 1 to 2 weeks following ovulation.

   b 2 to 3 days following ovulation.

   c 24 to 36 hours following ovulation.

   d at fertilization.

   e before fertilization.

20 The greatest adjustment to life outside the mother that must be made by the   _____

   neonate is:

   a regulation of the heart.

   b temperature regulation.

   c breathing air.

   d feeding.

   e defensive.

21 Which of the following is exchanged between fetal and maternal circulation?   _____

   a oxygen

   b carbon dioxide

   c blood

   d a and b

   e a, b, and c

22 Matching (answers must be used more than once).   _____

   a ___ skin   [1] endoderm

   b ___ heart   [2] mesoderm

   c ___ brain   [3] ectoderm

   d ___ skeletal muscles

23 A human embryo is most susceptible to the effects of teratogens during the first   _____

   two months of pregnancy.

   a true

   b false

24 The technique of obtaining cells from an unborn fetus by inserting a hypodermic needle through the abdominal wall of the mother into the amnion is called _____ .   _____

25 The use of high-frequency sound waves to locate and examine the fetus is termed _____ .   _____

Pharmacological Abbreviations

Abbreviation	Derivation	Meaning
ac	*ante cibum*	before meals
ad lib	*ad libitum*	as desired
aq	*aqua*	water
bid	*bis in die*	two times a day
bin	*bis in noctis*	two times a night
c	*cum*	with
caps	*capsula*	capsule
comp	*compositus*	compound
dil	*dilutus*	dilute
FDA		Food and Drug Administration
gm	*gramme*	gram
gr	*granum*	grain
gt, gtt	*gutta*	drop(s)
h	*hora*	hour
hs	*hora sommi*	bedtime
IM		intramuscular
IV		intravenous
mg		milligram
NPO	*nulla per os*	nothing by mouth
od	*omni die*	every day
oh	*omni hora*	every hour
ol	*oleum*	oil
om	*omni mane*	every morning
on	*omni nocte*	every night
os	*os*	mouth
oz		ounce
pc	*post cibum*	after meals
PDR		*Physician's Desk Reference*
po	*per os*	by mouth
prn	*pro re nata*	when requested
Q (q)	*quaque*	every
qd	*quaque die*	every day
qh	*quaque hora*	every hour
qid	*quater in die*	four times a day
qsuff	*quantum sufficit*	as much as sufficient
s	*sine*	without
sos	*si opus sit*	if necessary
subq		subcutaneous
syr	*syrupus*	syrup
tab		tablet
tid	*ter in die*	three times a day
ung	*ungentum*	ointment
U.S.P.		United States Pharmacopeia

Answers to the Post-Tests

---

1 distal; proximal   *2* atom, molecule, cell, tissue, organ, system, organism   *3a* 3
3b 4   *3c* 2   *3d* 1   *4* homeostasis   *5* a   *6* b   *7* system   *8* e   *9* e   *10* e
11a 4   *11b* 1   *11c* 2   *11d* 5   *11e* 3   *12* frontal   *13* viscera
14a pericardium   *14b* pleural   *15* membranes   *16* b   *17* b

CHAPTER 1

1 c   *2* a   *3* b   *4* auricular   *5* a   *6* bridge   *7* c   *8* thyroid   *9* b   *10* a
11 a   *12* second   *13* cubital   *14* linea alba   *15* a   *16* e   *17* b   *18a* 5
18b 2   *18c* 1   *18d* 4   *18e* 3   *19a* 2   *19b* 4   *19c* 3   *19d* 1   *19e* 5
20 McBurney's   *21* philtrum   *22* shin   *23* palpation   *24* a   *25* b

CHAPTER 2

1 a   *2* e   *3* a   *4* cytoplasm, nucleoplasm, plasma membrane   *5* b   *6* a
7a 2   *7b* 4   *7c* 3   *7d* 1   *8* cytoskeleton   *9* vacuoles   *10* nucleolus
11 cytosol   *12* a   *13* b   *14* cytoskeleton

CHAPTER 3

CHAPTER 4

1

	X		X
X		X	
	X		
X			
X			

2

X			
	X		
		X	
			X
X			

3 a   *4* desmosomes, tight junctions, gap junctions   *5* b   *6* c   *7* e   *8* mucus
9 adipose   *10* a   *11* a   *12* b   *13* b   *14a* 3   *14b* 4   *14c* 2   *14d* 1
15a 1   *15b* 4   *15c* 3   *15d* 2   *16* In the linings of the various body tubes that open
to the outside of the body.   *17* d   *18* d   *19* a   *20* b   *21* a   *22* a   *23* e
24 a   *25* e

1 It consists of tissue that performs specific activities.   *2* a   *3* sweat, sebaceous
4 c   *5* b   *6* b   *7a* 2   *7b* 3   *7c* 1   *7d* 4   *8* e   *9a* i, ii, iii
9b iii, i, ii   *9c* i, ii, iii   *9d* ii, iii, i   *10* c   *11* c   *12* a   *13* a   *14* matrix
15 c   *16* dermal   *17* decubitus ulcer   *18* papilloma viruses   *19* a   *20* c
21 c   *22* a   *23* sex hormone levels or levels of testosterone and estrogen
24 melanin   *25* blackheads

CHAPTER 5

X	X
X	X
X	X
	X
X	
X	X
	X
X	X
X	
1 X	X

2 a **3** d **4** a **5** d **6** e **7** a **8** e **9a** 1 **9b** 3 **9c** 4 **9d** 2 **10** a
11 epiphyses **12** diaphyses **13** e **14** d **15** a **16** lacuna **17** periosteum
18a 2 **18b** 3 **18c** 4 **18d** 1 **19** osteoblasts **20** b **21** internal composition,
shape **22a** osteogenic cells **22b** osteoblasts **22c** 2 **22d** 3 **23a** osteogenic
cells **23b** osteoblasts **23c** osteocytes **23d** osteoclasts **23e** bone-
lining cells **24a** calcium **24b** phosphate **24c** magnesium **25** myeloid

1 c **2** a **3** b **4** a **5** foramen **6** a **7** b **8** b **9a** cervical
9b thoracic **9c** lumbar **9d** sacral **9e** coccygeal **10a** 3 **10b** 8 **10c** 1
10d 6 **10e** 10 **10f** 5 **10g** 7 **10h** 9 **10i** 11 **10j** 2 **10k** 4 **10l** 12
11 c **12** a **13** d **14** b **15** b **16** e **17** a **18** c

1 a **2a** yes **2b** yes **2c** yes **2d** yes **2e** no **2f** no **2g** yes **2h** no
3a larger **3b** pointed **3c** oval **3d** shallow **4** a **5a** 7 **5b** 1 **5c** 8
5d 2 **5e** 9 **5f** 6 **5g** 11 **5h** 5 **5i** 10 **5j** 4 **5k** 12 **5l** 13 **5m** 14
5n 15 **5o** 3 **5p** 16 **6** b **7** c

1 e **2** e **3** e **4a** KJ **4b** N **4c** KJ **4d** HJ **4e** VC **4f** VC **5a** 1
5b 2 **5c** 5 **5d** 4 **5e** 3 **6** b **7** a **8** b **9** a **10** a **11** b
12 elbow, knee, ankle, or interphalangeal **13** synovial **14** b **15a** 3 **15b** 2
15c 5 **15d** 7 **15e** 9 **15f** 10 **15g** 8 **15h** 1 **15i** 6 **15j** 4 **16** b
17 a **18** fibrous **19** e **20** d **21** a **22** b **23** a
24 synovial membrane **25** luxation or dislocation

1 d **2** e **3** a **4** a **5** e **6a** 3 **6b** 4 **6c** 1 **6d** 8 **6e** 7 **6f** 9
6g 10 **6h** 5 **6i** 2 **6j** 6 **7** a **8** b **9** b **10** motor unit **11** I, H **12** e
13 e **14** a **15** single **16** heat **17** fascia

1 e   *2* origin   *3* e   *4* a   *5* b   *6* e   *7* c   *8* a   *9* b   *10* a   *11a* 3
11b 5   *11c* 1   *11d* 2   *11e* 4   *11f* 1   *11g* 5   *11h* 2   *11i* 4   *11j* 3
11k 2   *11l* 1   *11m* 1   *11n* 2   *11o* 3   *11p* 3   *11q* 1   *11r* 4   *11s* 2
11t 5   *11u* 2   *11v* 1   *11w* 3   *11x* 5   *11y* 4   *12* d   *13* c   *14a* 5   *14b* 4
14c 3   *14d* 2   *14e* 1   *14f* 3   *14g* 1   *14h* 4   *14i* 5   *14k* 2   *14l* 4
14m 1   *14n* 5   *14o* 3   *15* a   *16* synergists   *17* first   *18* a   *19* b   *20a* 5
20b 3   *20c* 2   *20d* 4   *20e* 1   *21a* 7   *21b* 2   *21c* 8   *21d* 3   *21e* 1
21f 4   *21g* 5   *21h* 6   *22a* 3   *22b* 1   *22c* 4   *22d* 2   *22e* 5   *23* e
24 a   *25* b

CHAPTER 11

1 e   *2* c   *3* e   *4* b   *5* b   *6* b   *7a* 5   *7b* 6   *7c* 8   *7d* 9   *7e* 1   *7f* 4
7g 10   *7h* 2   *7i* 7   *7j* 3   *8a* 2   *8b* 4   *8c* 6   *8d* 8   *8e* 10   *8f* 9
8g 1   *8h* 7   *8i* 5   *8j* 3   *9* e   *10* c   *11* somatic, vixceral
12 excitability, conductivity   *13* general somatic afferent, general visceral afferent, general somatic efferent, general visceral efferent, special afferent   *14* multipolar, bipolar, unipolar   *15* receptive, initial, conductive, transmissive   *16* a
17 all-or-none   *18* saltatory conduction   *19* Parkinson's disease, Huntington's chorea
20 divergent, convergent, feedback circuits, parallel circuits, two-neuron circuits, three-neuron circuits   *21* a   *22* e   *23* e   *24* b   *25* a

CHAPTER 12

1 e   *2* b   *3* a   *4* in the muscules of the thigh   *5* below L2   *6* a   *7* fascicles
8 lumbar   *9* e   *10* d   *11* e   *12* e   *13* e   *14* e   *15a* 5   *15b* 7   *15c* 1
15d 8   *15e* 4   *15f* 2   *15g* 3   *15h* 10   *15i* 9   *15j* 6   *16a* 3   *16b* 2
16c 4   *16d* 6   *16e* 9   *16f* 10   *16g* 8   *16h* 7   *16i* 1   *16j* 5   *17* a   *18* b
19 spinal shock   *20* 8   *21* b   *22* sciatica   *23* paraplegia   *24* epineurium, perineurium, endoneurium   *25* pyramidal, extrapyramidal

CHAPTER 13

1 e   *2* d   *3* c   *4* d   *5* c   *6* a   *7* a   *8* c   *9* c   *10* b   *11* a f g h d f b c e
12 hypothalamus   *13* right cerebellum   *14* See Figure 14.16 (pp. 374–375)   *15a* 2
15b 1   *15c* 4   *15d* 3   *16a* 5   *16b* 10   *16c* 9   *16d* 2   *16e* e   *16f* 10
16g 8   *16h* 7   *16i* 1   *16j* 5   *17i* b   *17ii* a   *17iii* a   *17iv* b   *17v* a
17vi b   *17vii* b   *17viii* a   *17ix* a   *17x* b   *18* midbrain   *19* theta   *20* L-dopa
21 epilepsy   *22* parietal   *23* regulation of reproductive cycles, blood-sugar levels, body fluid balance, metabolism, body temperature   *24* limbic system   *25* arbor vitae

CHAPTER 14

1a nicotinic   *1b* parasympathetic   *1c* sweat glands, blood vessels   *2* a   *3* d
4 e   *5* b   *6* c   *7* a   *8* a   *9* b   *10* biofeedback   *11* adrenergic
12 two   *13* collateral   *14* a   *15a* 1   *15b* 4   *15c* 3   *15d* 2
16 homeostasis   *17* a   *18* hypothalamus   *19* limbic   *20a* afferent receptor
20b afferent neuron   *20c* interneurons   *20d* two efferent neurons
20e visceral effector   *21a* brainstem   *21b* a reticular formation   *21c* spinal cord
21d hypothalamus   *21e* cerebral cortex   *21f* limbic system   *22* cholinergic
23 paravertebral (lateral), prevertebral (collateral), terminal (peripheral), sympathetic
24 visceral efferent system   *25* a

CHAPTER 15

1 d   *2* c   *3* e   *4* d   *5* c   *6* a   *7* b   *8* transducer   *9* mechanoreceptor
10 neuromuscular   *11* ossicles   *12* cochlea   *13* Corti   *14* hair   *15a* 2
15b 3   *15c* 1   *16a* 2   *16b* 3   *16c* 1   *17a* 2   *17b* 1   *17c* 3   *18a* 2
18b 1   *18c* 3   *19* e   *20* a   *21* b   *22* d   *23* b   *24* a   *25* otitis media

CHAPTER 16

1 e   *2* e   *3* e   *4* a   *5* endocrine   *6* thyroxin   *7* thyrotropin (TSH)
8 cortex   *9* adrenaline   *10a* 2   *10b* 1   *10c* 3   *11a* 2   *11b* 3   *11c* 1
12a 2   *12b* 3   *12c* 1   *13a* 1   *13b* 2   *13c* 3

1 e   *2* e   *3* c   *4* d   *5* c   *6* d   *7* c   *8* d   *9* b   *10* a   *11* b   *12* a
13 d   *14* e   *15* b   *16* d   *17* a

1 e   *2* e   *3* e   *4* d   *5* e   *6* a   *7* a   *8* e   *9* b   *10* b   *11* e   *12* c
13 b   *14* d   *15* d   *16* e   *17* b   *18* a   *19* a   *20* a   *21* a   *22* b   *23* b

1 b   *2a* 4   *2b* 2   *2c* 5   *2d* 1   *2e* 3   *3* a   *4* carotid sinus, aortic arch
5 hypertension   *6* infarction   *7* stroke   *8* brachial artery   *9* internal carotid
artery   *10* a   *11* d   *12* e   *13* endothelial cells   *14* tunica media, adventitia
15i b   *15ii* a   *15iii* b   *15iv* a   *16* sphygmomanometer   *17* cerebral arterial
circle (circle of Willis)   *18* inferior vena cava   *19* b   *20* d   *21* d   *22* a
23 d   *24a* 3   *24b* 4   *24c* 2   *24d* 5   *24e* 1   *25a* 3   *25b* 4   *25c* 2
25d 5   *25e* 1

1 b   *2* b   *3* spleen   *4* palatine, pharyngeal, lingual   *5* chemotaxis   *6* e   *7* b
8 b   *9* e   *10* c

1 alveoli   *2* nasal septum   *3* c   *4* b   *5* b   *6* a   *7* trachea   *8a* 2   *8b* 3
8c 4   *8d* 5   *8e* 1   *9* b   *10* d   *11* b   *12* b   *13* cartilage
14 external respiration   *15* internal respiration   *16* e   *17i* a   *17ii* b   *17iii* a
17iv b   *17v* b

1a 10   *1b* 7   *1c* 1   *1d* 5   *1e* 3   *1f* 2   *1g* 9   *1h* 4   *1i* 8   *1j* 6   *2* d
3 b   *4* pyloric sphincter   *5* parietal   *6* pepsinogen   *7* bile, gallbladder
8i a   *8ii* b   *8iii* a   *8iv* b   *8v* b   *9* b   *10* lesser omentum
11 taeniae coli, haustra   *12* e   *13* c   *14* a   *15* e   *16* d   *17i* b   *17ii* a
17iii b   *17iv* b   *17v* a   *18* uvula   *19* a   *20* c   *21* molars   *22* rugae
23 d   *24* b   *25* hepatocytes

1a 7   *1b* 3   *1c* 6   *1d* 2   *1e* 4   *1f* 5   *1g* 1   *2* d   *3* e   *4* c   *5a* H_2O
5b Na^+ *5c* Ca^{2+} *5d* K^+ *5e* H^+ *6* 180   *7* one liter, two liters   *8* e   *9* a
10 ureter   *11* b   *12* glomerular (Bowman's) capsule   *13* b   *14* e   *15* d   *16* a

1a primary spermatocyte, secondary spermatocyte   *1b* seminiferous tubules
1c orgasm   *2i* b   *2ii* a   *2iii* b   *2iv* a   *2v* a   *3* e   *4* hyaluronidase, ovum
5 d   *6* e   *7* e   *8* a   *9* d   *10* c   *11* e   *12* b   *13a* 2   *13b* 5   *13c* 4
13d 3   *13e* 1   *14a* 2   *14b* 5   *14c* 1   *14d* 3   *14e* 4   *15a* 5   *15b* 4
15c 3   *15d* 2   *15e* 1   *16a* 2   *16b* 4   *16c* 3   *16d* 1   *17a* 1   *17b* 2
17c 3   *17d* 4   *18* e   *19* a   *20* b   *21* b

CHAPTER 25

1a 3 *1b* 1 *1c* 4 *1d* 5 *1e* 2 *2* fetus *3* yolk sac *4* zygote *5* d
6 d *7* b *8* meconium *9i* a *9ii* b *9iii* a *9iv* b *9v* b *10* b
11 zygote morula blastocyst embryo fetus *12i* b *12ii* b *12iii* b *12iv* b
12v a *13* c *14* b *15* seven *16* sixth *17* e *18* d *19* a *20* c
21 d *22a* 3 *22b* 2 *22c* 3 *22d* 1 *22e* 2 *23* a *24* amniocentesis
25 ultrasonography

CHAPTER 26